RCS Synthesis for Chipless RFID

Remote Identification Beyond RFID Set

coordinated by
Etienne Perret

RCS Synthesis for Chipless RFID

Theory and Design

Olivier Rance
Etienne Perret
Romain Siragusa
Pierre Lemaître-Auger

ELSEVIER

First published 2017 in Great Britain and the United States by ISTE Press Ltd and Elsevier Ltd

ISTE Press Ltd
27-37 St George's Road
London SW19 4EU
UK

www.iste.co.uk

Elsevier Ltd
The Boulevard, Langford Lane
Kidlington, Oxford, OX5 1GB
UK

www.elsevier.com

Notices

MATLAB$^{®}$ is a trademark of The MathWorks, Inc. and is used with permission. The MathWorks does not warrant the accuracy of the text or exercises in this book. This book's use or discussion of MATLAB$^{®}$ software or related products does not constitute endorsement or sponsorship by The MathWorks of a particular pedagogical approach or particular use of the MATLAB$^{®}$ software.

For information on all our publications visit our website at http://store.elsevier.com/

British Library Cataloguing-in-Publication Data
A CIP record for this book is available from the British Library
Library of Congress Cataloging in Publication Data
A catalog record for this book is available from the Library of Congress
ISBN 978-1-78548-144-4

Printed and bound in the UK and US

Contents

Introduction

Automatic identification technologies have profoundly changed the consumption patterns and organization of businesses. They have enabled the use of merchandise monitoring systems that would not otherwise be possible. Barcodes are the most successful example of identification technology. Invented in the 1950s and used industrially since the 1970s, barcodes have become essential for the field of large-scale distribution. Today, barcodes are featured on more than 70% of manufactured objects around the world. The success of barcodes can be attributed to their very low cost and great ease of use.

In the 1990s, technologies based on the use of electromagnetic waves also appeared on the identification market. Radio Frequency Identification (RFID) has been able to hold its ground against barcodes in some fields, such as logistics, tracking and access control, by offering new possibilities. RFID added functionalities like remote reading, batch reading and the ability to modify information contained in the tag. However, RFID has not replaced barcodes in large-scale distribution, where the price of the tag is sometimes comparable to the goods it identifies.

The growth of RFID is hindered by the unit cost of tags. For a few years now, research has been focused on designing RFID tags that do not contain silicon microchips. Chipless RFID is situated, from an application standpoint, at the intersection between barcodes and traditional RFID. Without a chip, the identification of the tag is not contained in a memory, as in the case of traditional RFID, but directly in the geometry of the metal pattern. In this sense, chipless tags can be likened to radar targets designed to have specific and easily identifiable electromagnetic signatures. For this type of device, the amount of information that it can store is central to the question of design, especially because tags are limited by their compactness. The quantity of information offered by current coding techniques is not sufficient for industrial use. This is a technological barrier that must be resolved before chipless RFID can become a real alternative to barcodes or traditional RFID.

In response to this issue, this book proposes a new coding method based on the general form of the tag's signature. To do this, it must be possible to make tags whose signature is given in advance, which creates a complex inverse problem. A design method based on the assembly of more or less resonant metal patterns is proposed. This approach essentially amounts to projecting the desired signature onto the space formed by the resonators whose response is assumed to be known. The response of the unit cell will be controlled through geometric manipulations within the tag.

This book consists of five chapters:

– Chapter 1 is an introduction that will situate chipless RFID in relation to other major identification technologies. We will show that chipless RFID is a kind of compromise between barcodes and traditional RFID in terms of performance as well as application.

– Chapter 2 is an overview of the coding methods used in chipless RFID. The different approaches (temporal, frequency and hybrid) are explained and illustrated using examples from the literature. It will be established that the greatest amount of information is obtained by tags coded using the frequency approach.

– Chapter 3 presents the different physical principles involved when a chipless tag is interrogated. Theoretical concepts from different fields such as radar, antennas, resonant systems and traditional RFID are compiled and provide tools for the design and analysis of chipless RFID tags.

– Chapter 4 presents the concept of amplitude coding. Two examples of tags with different configurations (one with ground plane and one without) will be designed. An assessment procedure will be presented in order to practically evaluate the coding performances attainable for each configuration. Amplitude coding is an important preliminary step to coding on the general form of the tag. The techniques used to control the amplitude of the tags' response will be revisited in the following chapter.

– Chapter 5 presents the design method for tags whose coding is based on the general form of the signature. Two different cases will be studied. The first case includes resonant unit cells having a ground plane and the second case includes less-resonant structures without a ground plane. We will see that in the broadband case, the primary difficulty is related to couplings that appear between the different resonators.

In the Conclusion, we will summarize the results obtained in this study and discuss the new perspectives that can arise from them.

Automatic Identification Technology

All commercial or industrial activity requires a tracking system that makes it possible to check inventory, input-output and product consumption, circulation of documents, materials, equipment, etc. This monitoring can be carried out manually by keeping records or by digital processing. In the latter case, information must be entered into a computer. Data entry by a human operator has a number of disadvantages. The error rate is relatively high, in the order of 2–3% of typed entries. Entry speed is often low because it is done by people for whom data entry is just another task on top of their other work. These elements make it so that keyboard input is often considered too costly and cumbersome to be used.

Automatic identification is a set of techniques, primarily including barcodes and RFID, that are intended to automate data entry. The speed and security of the input has several advantages over manual entry and enables the implementation of an information system that would not be possible otherwise. In companies, automatic identification is used in almost all areas: merchandise receiving, inventory, operation reporting, quality control, order preparation and product processing, etc.

In this chapter, we will present the two main automatic identification technologies, barcodes and RFID (and in particular, passive UHF technology). We will demonstrate that on the one hand, barcodes are limited in terms of functionality and on the other, the price of RFID is prohibitive for a large number of applications. We will also introduce a bridge technology that is still in its research stage that makes it possible to combine certain functions of RFID with the very low cost of barcodes: chipless RFID. Chipless RFID is the main subject of this book. This chapter will introduce the reader to the issues and philosophy related to this technology. The technical and scientific aspects will be presented more in-depth in the following chapters.

1.1. Barcodes

An identification system based on barcodes is made up of a label that is generally printed using a thermal printer with an optical scanner. The original patent [WOO 52] was filed by two American students in 1952 for a school project to develop a method to automate product registration for manufacturers. By the end of the 1970s, barcodes had become a commercial success thanks to the adoption of the Universal Product Code (UPC Code) in large-scale distribution. Since then the use of barcodes has extended to several other areas, including tracking mail and airline luggage, identifying medication, indexing documents, etc. Today, barcodes are the most common automatic identification solution, used on 70% of all manufactured objects [PAL 91].

The predominance of barcodes over other automatic identification techniques can be explained for the most part by their low production cost and their ease of printing and use [PAL 91]. The unit cost to implement a barcode is estimated at about 0.005 USD. Barcodes are also very reliable, with an error probability of one in 2 million. Significant developments in readers have allowed the barcode to remain competitive, including an increase in read range and the possibility of in-flight reading [MAC 89]. Finally, barcodes benefit from standards adopted globally, which for a long time gave them a significant advantage over younger technologies like RFID.

1.1.1. *Labels*

A barcode (Figure 1.1(a)) is composed of a series of vertical lines of varying widths (called bars) and spaces. The characters are encoded using a combination of bars and spaces. Since 1999, 2D barcodes have also appeared on the automatic identification market (Figure 1.1(b)). Although 2D barcodes use pixels instead of bars to code the characters, the general idea remains identical to 1D barcodes.

Figure 1.1. *Structure of barcodes. a) 1D barcode using Code 39;*
b) 2D barcode using the QR code

The transposition of characters into barcode form is called symbology. This involves coding as well as the use of markers to indicate the beginning and end of the information. Since their appearance, many codes have been designed to improve the consistency, reliability, ease of reading and printing of barcodes. In principle, the old codes are replaced by more recent codes that perform better. However, some industries established standards based on old barcodes, which are still used in those sectors. For example, the code EAN 13 (Figure 1.2) is used all over the world on consumer goods despite its relatively low 43-bit coding capacity. Contrary to popular belief, barcodes generally do not contain any descriptive data such as price or a description of the item. The data in a barcode represents only a reference used by the computer to carry out the search of a database.

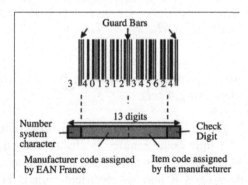

Figure 1.2. *Representation of EAN 13 barcodes used in large-scale distribution. It makes it possible to code 13 digits (i.e. 43 bits). For a color version of this figure, see www.iste.co.uk/rance/rfid.zip*

There are many types of barcodes. The most common barcodes are compared in Table 1.1.

	EAN 13	Code 39	Code 128	Data matrix
Date	1970	1974	1981	1999
Encoded characters	Numerical (digits 0 to 9)	Alphanumeric (0 to 9, A to Z, special characters) 43 characters	Complete ASCII (128 characters)	Alphanumeric characters
Identifier coding capacity (varies by size)	Fixed capacity: 13 digits (43 bits)	130 bits for 24 characters	168 bits for 24 characters	862 characters (4.5 kbit) for a tag that is 88 mm × 88 mm
Identifier length	Varies between 20 mm and 37 mm according to printing conditions	95 mm for 24 characters[1]	75 mm for 24 characters[1]	Varies between 10 mm × 10 mm and 144 mm × 144 mm
Properties	Fixed length, Common code in 60 countries	Universal code, read by most equipment	Very dense code but only read by certain readers	Very large coding capacity but requires specific readers
Example of application	Food products, large-scale distribution items	Product management and sales, labeling, distribution, document indexing	Transport, industry, health, distribution	Quick link to online content, exchange payment information, tracking batches or items
Reading device	All barcode readers	Barcode readers	Barcode readers	Smartphones, Imagers

Table 1.1. *Comparison between different types of common barcodes*

1.1.2. *Different types of readers*

Classic barcodes owe their longevity to innovations made in optical reading. Today, there are three types of readers: laser readers, CCD (charge-coupled device) readers and imagers.

Laser readers (Figure 1.3) rely on an optical reflection method. When reading a barcode, the light beam emitted is absorbed by the dark bars and reflected by the light spaces. Within the reader, a phototransistor receives the reflected light and converts it into an electrical signal. The length of the

---

1 Printing of 24 data characters, with a thermal printer making straight 0.25 mm bars.

electrical signal determines whether the elements are wide or straight. The reader is also equipped with a decoder that enables it to complete the transposition between the signal and the characters represented by the barcode. The laser reader uses a single ray of light generated by a laser diode. The light source is dense and precise, which allows for reading at close range or a few meters away, as well as on-the-fly reading, of objects or documents in movement.

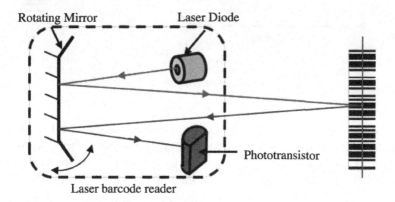

Figure 1.3. *Operating principle of a laser barcode reader. For a color version of this figure, see www.iste.co.uk/rance/rfid.zip*

Current laser readers automatically read code. There is no need to scan the entire length of the code, because a motorized mirror does that by reflecting the laser beam across the code, giving the illusion of a solid line. Some readers carry out this scanning on the height of the code (multiframe reading), and others multiply the scan to create a grid on which the code can be placed in any direction (omnidirectional reading). For some applications, especially in warehouses (superior lasers), some laser readers are capable of reading from a distance of more than 10 m.

CCD readers (Figure 1.4) function on the same principle as laser readers but they emit a scattered beam generated by a diode array, which makes it possible to accentuate the contrast between thin bars. The reflected image is focused through a lens and sent through a CCD sensor (the same type of sensors used in film and digital cameras).

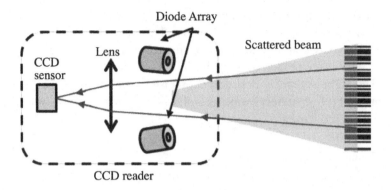

Figure 1.4. *Operating principle of a CDD reader. For a color version of this figure, see www.iste.co.uk/rance/rfid.zip*

A CCD reader allows barcodes to be read automatically: it's not necessary to sweep them across the scanner. The read range varies and depends on the reader's settings as well as the density of the barcodes to read. The greater the distance, the less the code is illuminated. CCD models do not have mechanical components and are generally very durable. Low-end CCD readers (commonly called "triggerless scanners") are generally the most cost-effective products. This technology has made huge strides in the past few years and read range has increased from 3 cm to 2 m for some top-of-the-line CCD readers.

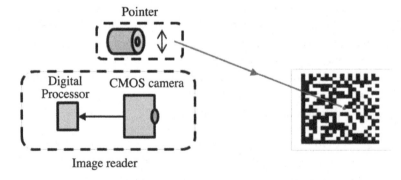

Figure 1.5. *Operating principle of an imager. An imager is a reader that is specially designed to read 2D tags. For a color version of this figure, see www.iste.co.uk/rance/rfid.zip*

Imagers, or 2D readers (see Figure 1.5), use a different technology than laser and CCD readers. They use a CMOS camera that takes a photo of the code which they then analyze and decode. In order to be read, 2D barcodes must be scanned vertically and horizontally.

The imager has good results with barcodes that are difficult to read or are damaged. It is capable of reading barcodes from any angle (omnidirectional reading). Finally, imagers do not contain any fragile or mobile components. This means that they are perfectly adapted for industrial environments like the automotive sector. Imagers have a maximum read range of around 30 cm.

Today, most barcode readers are wireless and communicate with a computer through an RF communication link. We can distinguish "readers", which read and transmit information in real time, from "terminals", which can store information and transmit them at a later date. Terminals are generally equipped with embedded operating systems, thus making it possible to upload dedicated software and applications.

The advantages and disadvantages of the different types of readers are compared in Table 1.2. As a guide, the number one seller in the "barcode reader" category on Amazon in 2016 was the BCST-20 developed by Inateck. The BCST-20 is a laser reader, with a nominal read range from 6–30 cm, and costs 45 euros.

	Laser Readers	*CCD Readers*	*Imagers*
Advantages	– Long-range reading – Large read range – Suitable for moving objects	– Compact – Inexpensive – Long lifespan due to lack of motor	– Versatile reader than can read all types of codes (1D, 2D) – Quick reading
Disadvantages	– Expensive	– Limited read range – Not suitable for moving objects	– Limited read range – Not suitable for moving objects – Expensive

Table 1.2. *Comparison of the advantages of different types of barcode readers*

Despite being very effective for their low cost, barcodes are less manageable in terms of reading (direct view and limited range for low-cost readers) and cannot have supplementary functions such as write access or sensor integration added to them. For some fields of application (tracking, logistics, etc.), these limitations have led to the appearance of alternative solutions with greater operating ranges such as RFID.

1.2. RFID

1.2.1. *General introduction*

The second option for the implementation of an automatic identification system is to use the Radio Frequency Identification technology, commonly called RFID. An RFID system (Figure 1.6) is composed of a reader, generally connected to a database, that communicates with a label (or tag) via a radio frequency link.

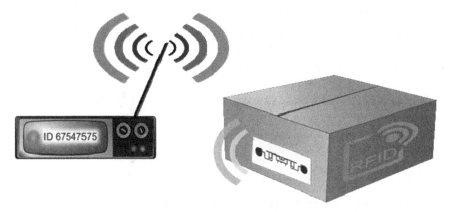

Figure 1.6. *General presentation of an RFID system. The reader communicates with a tag via an RF link. The tag is composed of a chip (red dot) connected to an antenna. For a color version of this figure, see www.iste.co.uk/rance/rfid.zip*

The tag is composed of an integrated circuit (chip) connected to an antenna. For miniaturized versions, it is possible to include this circuit on a paper label to create smart labels. In other cases, the circuit is embedded in plastic and the RFID tag can be presented in various forms (button, plastic card, etc.) or even included directly in a molded object (palettes, containers, vehicle keys, identification badges, etc.).

RFID tags have three properties that open up a wider field of use than barcodes. Firstly, it is possible to read RFID tags without a direct line of sight between the reader and the tag (for example, they can be read inside a box, through a layer of paint, or when they are molded in plastic, etc.). This disrupts the visual integrity of the product less. Secondly, it is possible to not only read the data, but also write it. Thirdly, anti-collision systems make it possible to read a large number of tags in the same volume almost simultaneously (reading the identifiers of all objects located in a specific given environment).

Read ranges depend on the technologies used (frequencies, type and antenna dimensions) and the available power, which varies according to local regulations. Generally, read ranges are between a few centimeters to a few meters. Distances of several hundreds of meters are technically possible, but these are non-standard applications (power, etc.) and/or using active tags that have their own power source (much more costly and larger than passive tags, such as automatic payment badges).

1.2.2. *History*

The original patent was filed by E. Brard in 1930 [BRA 30]. The principle of RFID was used in 1940 during World War II to identify aircrafts (*IFF: Identification Friend or Foe*). It consisted of completing the RADAR signatures of planes by reading a fixed identifier that enabled the authentication of allied planes [BRO 99]. During the 1960s and 1970s, RFID systems remained a classified military technology used to control access to sensitive sites [KOE 75], especially those related to nuclear energy.

The first general public application of RFID appeared in the 1970s with EAS (Electronic Article Surveillance) anti-theft systems whose transponders were the equivalent of 1-bit chipless tags. It was a commercial success and generated interest in RFID for large companies like General Electric and Philips.

In 1980, technological advances (transistors, miniaturization of components) allowed for the implementation of the passive tag, whose concept had been known [STO 48] and studied [HAR 63] since 1948. The RFID tag backscattered the wave received from the reader to transmit information. This technology made it possible to move away from having an energy source embedded in the tag, thereby reducing its cost and maintenance. From that point on, there was a great deal of enthusiasm for RFID, slated as the technology of the future. However, neither the frequencies nor the information exchange protocols were standardized, and consequently, the only possible applications were limited or internal to companies.

It was only in the 1990s that standardization of RFID equipment began and RFID took off. We witnessed the rollout of applications to many fields: "contactless" identification badges, public transportation (RATP), anti-theft systems (car keys with an RFID chip recognized by the vehicle and authorizing start-up), highway tolls (with active labels), etc.

In 1999, MIT (Massachusetts Institute of Technology) founded a research center specialized in the automatic identification (primarily RFID) called the Auto-ID Center. In 2004, Auto-ID became "EPCglobal," an organization charged with promoting the EPC (Electronic Product Code) standard, extending from the barcode to RFID. The ISO (International Standard Organization) also greatly contributed to the implementation of technical and application systems allowing for a high degree of interoperability and interchangeability.

In France, the CNRFID (Centre National de Référence RFID) was created by the Ministry of Economy, Industry and Digital Technology in 2009. This association aims to facilitate the adoption and appropriation of contactless technologies by companies.

Today, RFID technologies are very widespread in almost all industry sectors (aeronautics, automobiles, logistics, transport, health, etc.) as well as in daily life. Some examples of recent applications are the appearance of the contactless payment card and the massive rollout of RFID on the production line at Décathlon.

1.2.3. Classification of RFID tags

As previously noted, there are a wide variety of applications requiring remote object identification. These applications are subject to varying constraints on read ranges, the nature of objects to identify (permittivity, presence of metal, etc.) and even the usage environment of the tags. This wide variety of applications and their specific restrictions in part explain the great diversity of RFID technologies.

RFID tags are classified according to different criteria. The first condition that determines the price of the tag is the presence or absence of a power source within the tag. Following this factor, we can distinguish three types of tags.

Active tags contain an energy source used for both powering the chip and sending the response via an RF transmitter stage. Communication with the reader is therefore peer-to-peer and it is possible to use different frequencies for transmitting and receiving. Active tags generally make it possible to reach greater read ranges but the presence of a battery means that they are larger, costlier and require maintenance.

Semi-passive tags or battery-assisted passive tags have an embedded energy source used only to power the logical part of the chip and not to transmit the signal responsible for the tag's response. This means that the tag backscatters the wave emitted by the reader for communication, which is different from active tags. This type of tag is widely used for applications that require information be captured (temperature, shock, light, etc.) independent of the presence of an interrogator.

Passive tags use the wave emitted by the reader both to power the chip and for communication. This type of tag functions the most similar to barcodes and it is by far the most common on the RFID market due to its low cost and lack of maintenance.

Passive RFID tags are by far the most used on the current market. The different types of passive tags are generally classified by the system's operating frequency. Depending on the frequency used, the physical principles implemented are not the same; this is therefore a determining factor for the read range and general performance of the tag. The general properties of passive tags depending on the frequency used are presented in Table 1.3.

As a comparison, the price of a barcode is estimated at 0.005 euros, which is a much lower cost than the solutions offered by traditional RFID.

Frequency	LF 125 and 134.2 kHZ	HF 13.56 MHz	UHF 868 to 915 MHz	SHF 2.45 and 5.8 GHz
Physical principle	Inductive coupling (near-field)	Inductive coupling (near-field)	Propagation (far-field)	Propagation (far-field)
Max. typical range	0.5 m	1 m	10 m	1 m
General properties	- Relatively expensive even in large volumes. - Antenna requires a large number of coils. - Slight decrease in performance when metals or liquids are present.	- Less expensive than LF tags. - Well-adapted for applications that do not require reading many tags at a great distance. -Unique frequency in the world.	- In large volumes, UHF tags are less expensive than HF and LF tags. - Adapted for reading large volumes at long distances. - Less effective performance than HF when metals or liquids are present.	- Similar performance to UHF. - Highly sensitive to metals and liquids. - More directive reader/tag liaison for lower frequencies.
Unit cost	> 1 euro	> 0.4 euro	> 0.2 euro	> 0.3 euro
Major standards	ISO 14223/1 ISO 18000-2	ISO 14443 ISO 15693 ISO 18000-3	ISO 18000-6	ISO 18000-4

Table 1.3. *Properties of passive RFID tags depending on the frequency used*

1.2.4. *The passive RFID market*

Due to its low cost and good read range, UHF RFID is the most promising option for labeling consumer items. The evolution of the distribution of the passive RFID market according to the frequency used for the years 2011, 2013 and 2020 (numerical data and predictions from [DAS 10]) is represented in Figure 1.7. In 2011, the two most-used technologies were HF RFID (46.5%) and UHF RFID (40.9%). However, HF technology is losing momentum ahead of UHF RFID, which has seen a rapid increase. The trend reversal was already evident in 2013, when there were 2,182 million tags sold for HF RFID (about 37% of the total market) compared to 3,079 million for UHF RFID (52%). According to the study

[DAS 10], the estimated number of passive RFID tags that will be sold on the market in 2020, including all applications and markets, will have the following distribution: 1,308 million for LF RFID (3.5%), 5,904 million for HF RFID (15.8%) and 30,000 million for UHF RFID (80.6%).

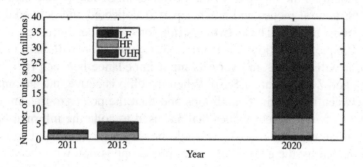

Figure 1.7. *Market evolution of passive RFID according to operating frequency. For a color version of this figure, see www.iste.co.uk/rance/rfid.zip*

1.2.5. *Passive UHF RFID tag function: backscatter method*

Because UHF RFID tags are currently the most common passive RFID tags, it is important to provide a few details about how they work. Passive UHF tags operate on a frequency between 860 and 960 MHz. At these frequencies, the preferred working mode of these RFID systems is the far-field. Greater read ranges can be reached compared to LF or HF inductive coupling tags.

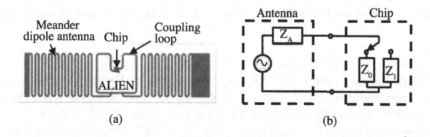

Figure 1.8. *a) Structure of a traditional UHF RFID tag; b) Equivalent electrical circuit*

A UHF RFID tag (Figure 1.8(a)) is made up of an antenna connected to an electronic chip that can both be modeled by complex impedances like in the equivalent electrical diagram presented in Figure 1.8(b). Communication relies on the concept of load modulation. When the wave transmitted by the reader encounters the tag, part of the power is immediately reflected by the antenna (structural mode) and another part is collected and delivered to the chip (antenna mode). These two distributions will be discussed in more detail in Chapter 3, section 3.3. If the power collected is sufficient to power the chip, it will activate and vary its input impedance between two distinct states (Z_0 and Z_1 in Figure 1.8(b)). When the chip is active, the antenna load varies between two distinct conditions and then the power re-transmitted by the tag takes two different values that are used to code the information. For example, in an extreme case, there could be an open-circuit load so that the signal reflected by the chip would be in phase opposition with the structural mode. The amplitude of the total reflected signal would therefore be the minimum, which can be used to code logic state "0". As a second example, there could be a short-circuit load so that the signal reflected by the chip would be in phase with the structural mode. The amplitude of the total reflected signal would therefore be the maximum, which can be used to code the logic state "1".

Two conditions must be satisfied to complete communication between the tag and a reader in a practical case. The first condition is to ensure that the power arriving at the chip is sufficiently large, which is to say greater than the minimum power required to activate the chip. The second condition is to ensure that the two coding states received by the reader are sufficiently differentiated. The difficulty of designing UHF RFID tags is due to the fact that these two conditions are often contradictory from the standpoint of antenna matching [PER 14]. A compromise must be found in order to optimize the read range of this system.

It is also important to consider how the chip works. Figure 1.9 shows a UHD RFID tag in the form of a functional diagram. The chip includes an analogous radiofrequency part (front-end radio) and a digital part. The digital part is usually composed of a state machine that has the role of analyzing the instructions received, coding or decoding the information and responding by sending these data to the front-end radio. The digital part is activated by the front-end radio when the power collected by the antenna is sufficient (activation power of the chip). The front-end radio has three main functions: energy harvesting, signal reception and load modulation.

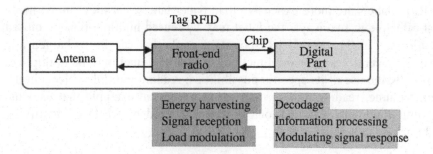

Figure 1.9. *Operational diagram of a passive UHF RFID tag. For a color version of this figure, see www.iste.co.uk/rance/rfid.zip*

Energy harvesting is generally ensured by a rectifier that makes it possible to retrieve a direct current voltage from the carrier received by the antenna in order to power the tag. The rectifier has a voltage limiter to protect the chip from surges. An output reservoir capacity in the rectifier ensures power to the chip during the backscatter phase.

The reception of data transmitted by the reader is ensured by a traditional receiver chain including a demodulator, a baseband filter and an analog-to-digital converter. Decoding and information processing are completed by the digital part.

The front-end RF also guarantees load modulation. In other words, it includes a device that makes it possible to modify the input impedance of the chip. This device is generally made up of a transistor controlled by the modulating digital signal that switches between two impedances with different values.

1.2.6. *Limitations of RFID*

The cost of chips, which is generally around 20 cents per unit, is the primary commercial obstacle of RFID technology. RFID chips equipped with a battery or a rewritable memory can cost up to 20 euros a piece. Changing equipment, purchasing new readers, and especially investing in adapted logistical infrastructure are all factors in the high cost, although they are one-time costs.

RFID labels are more expensive than paper labels, but they can be made cost-effective in two ways: the label may be reused multiple times ("closed circuit"), as in the case of transportation cards that can be recharged and used for several years, or the functions of a single label may be multiplied, especially the use of its specific possibilities: reading multiple objects in the same volume, reading/writing, and so on. A typical example that uses the maximum amount of the various functions offered by RFID is identifying and tracking pallets.

RFID is used successfully when the cost of the tag becomes negligible in relation to the value of the item that it identifies. The price of a UHF tag of around 20 cents is, for example, much lower than the value of a pallet of merchandise. However, to identify many objects in the large-scale distribution sector, the price of the tag is often non-negligible in relation to the price of the object to be identified. In this domain, barcodes remain by far the most common solution due to their very low cost, ease of use and reliability.

Another major disadvantage of RFID compared to barcodes is the wide variety of tags that exist on the market. In fact, there is no universal tag capable of responding to the needs of all applications and different technical solutions are proposed according to read ranges, environments, usages, etc. If the tag is not designed for the specific application, there is a significant risk that its performance will be greatly decreased. One illustrative example is the use of RFID tags to label products that contain metal, such as boxes of canned goods. If the tag is not specially designed for this application, reading is simply impossible. This profusion of tags makes access to RFID difficult for businesses and it is often necessary to ask a specialist to find out what type of tag is adapted for the desired application. On the other hand, barcodes are much simpler to access. The concept and its possibilities are well-known to the greater public and a simple Internet search is enough to determine what barcode is well-adapted to the desired use. The barcode label is easier to print directly with a computer and phasing in equipment is relatively simple.

1.3. Chipless RFID

RFID benefits from a very large operational spectrum compared to optical barcodes. The use of RF waves to communicate makes it possible to increase read ranges and detect a tag without a direct line of sight, regardless of its orientation. The presence of the chip adds the possibility of writing and allows for volumetric readings thanks to the use of communication protocols. These functions are impossible to implement with barcodes, and yet barcodes remain the main player in automatic identification technologies, equipping 70% of manufactured items. The success of barcodes can be explained very simply: they are the least expensive identification solution [PER 14].

In response to this observation, significant efforts have been made in recent years to reduce the cost of RFID labels. All the same, it seems that they will remain more expensive than barcode labels, which are only printed paper, which remain necessary, while RFID must add the cost of the chip and the antenna. It is also true that traditional RFID provides some very valuable functionalities, so the question is: is it possible to design a technology based on RF waves but capable of competing with barcodes in terms of price?

In the mid-2000s, researchers found an answer to this question [PER 14]. The only way to drastically reduce the cost to under the accepted threshold, estimated at 1 cent, is to create RFID labels that do not have a chip. Although it is still in the research stages, the concept of chipless RFID is a promising solution that makes it possible to combine certain functionalities of RFID with the low cost and simplicity of barcodes. Once it is fully developed, chipless RFID should be able to compete with barcodes in certain areas of application.

1.3.1. *General operating principle*

Like other identification technologies, a chipless RFID system is composed of an RFID label and a reader (Figure 1.10). The operating principle of the reader is very similar to that of radar. The reader transmits an RF wave in the direction of the tag and analyzes the form of the reflected signal to decode the information contained in the tag. Given that the communication medium is an RF wave, the system shares the ability to read a tag without a direct line of sight with traditional RFID. For example, the tag can be read when it is placed inside a box. A chipless tag is made up of a substrate on which there are conductive patterns that give the tag an easily recognizable specific signature. As in the case of barcodes, the label is completely passive and read-only. Similarly, the absence of a chip in the tag does not allow for the implementation of a communication protocol based on information sequencing. It is therefore more difficult to implement anti-collision methods than in the case of traditional RFID and the problem should be approached not by using a specific communication protocol, but as a physical problem of interference.

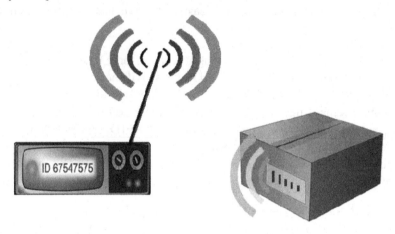

Figure 1.10. *General operating method of a chipless RFID system. For a color version of this figure, see www.iste.co.uk/rance/rfid.zip*

Although it partly reduces the operational field of a chipless tag compared to a traditional RFID tag, the absence of the chip has some advantages. First of all, and this is the main objective, the absence of a chip significantly reduces the price of the tag for mass market applications like barcodes. Compared to traditional RFID, it is possible to save the price of the chip as well as eliminate the step of connecting the chip and the antenna. What's more, the absence of a localized electronic component makes manufacturing compatible with a printing process if using a conductive ink. Research has shown that a flexographic printed tag would have a comparable cost to optical barcodes [VEN 13d]. Another advantage of chipless RFID lies in the fact that all the power received by the tag can be used for communication because there is no chip to power. Consequently, chipless systems generally operate with lower power levels than traditional RFID. Finally, the absence of a chip gives a chipless tag more operational reliability, a potentially longer life span, and thermal and mechanical resistances that are far superior to tags with chips [KIM 13].

In a traditional RFID tag, all information is contained in the chip. With chipless tags, the first question that comes to mind is to find out how the information is coded and how much information the tag is capable of storing. This issue is fundamental to chipless RFID because the quantity of information contained in the tags is generally low, in the order of a few tens of bits. This value is much lower than a traditional RFID tag or recent barcodes. In the case of UHF RFID, the amount of information can impact the price slightly or the reading time but will not modify the dimensions of the final label nor the communication frequency of the system. Similar to barcodes, the question of the quantity of information simply determines the type of code to use (classic or 2D codes) but no longer constitutes a technological barrier. In the case of chipless RFID, on the other hand, the quantity of information is a major challenge that will impact a large number of parameters, in particular the dimensions of the tags and the frequency range used.

The issue of the quantity of information is at the core of this book and we will come back to it throughout the study, in particular in the following chapter, which is a survey of coding methods used for chipless RFID.

1.3.2. Basic example of a chipless tag and performance factors

In order to illustrate how information is coded in a chipless tag, we will study an example of a very simple tag from [JAL 05], represented in Figure 1.11. This basic example will also allow us to identify the figures of merit that will be used in the next chapter to compare the performance of the tags.

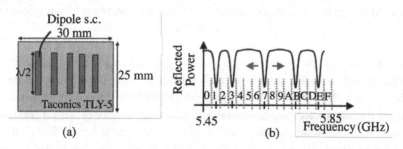

Figure 1.11. *a) Structure of a chipless RFID tag proposed in [JAL 05]; b) Spectral signature of the tag. Frequency coding method using positioning of dips. The identifier associated with the tag is 137BE in hexadecimals. For a color version of this figure, see www.iste.co.uk/rance/rfid.zip*

This tag is made up of five dipoles in short circuit of varying sizes arranged on a traditional RF substrate with a ground plane. The dipoles act as half-wave resonators. When an incident wave encounters the tag, a resonance phenomenon takes place for certain specific frequencies that correspond to the dipole resonances. These resonances are in phase opposition with the wideband response of the ground plane and translate into easily identifiable dips at the level of the tag's spectral signature (Figure 1.11(b)). The resonance frequency of the dipoles can be adjusted simply by modifying their length. In this way, we can adjust the position of the dips at the spectrum level and thus code information. For example, we can divide the frequency band in N windows of identical sizes that each correspond to a digit. The identifier of the tag is therefore determined by the digits that are associated with the windows where there is a dip. This very simple example will serve to illustrate the main figures of merit of chipless tags.

1.3.2.1. Coding capacity

As noted above, the most important figure of merit after the price of the tag is the quantity of information contained in the tag, known as the coding capacity. In this example, it is quite clear that the number of different configurations offered by the tag depends on both the number of resonators present in the tag (the surface area) and the number of frequency windows (the available frequency range).

1.3.2.2. Frequency range

One noteworthy point about the example given in Figure 1.11 is that, unlike a traditional RFID tag, the coding is carried out in the frequency domain. In order to code more information, we can increase the number of resonators and therefore consider a larger frequency range. Generally, the greater the band used, the greater the possibility of having tags with a large coding capacity. In fact, the capacity of the tag depends on the number of frequency windows associated with each resonator. The more windows there are, the greater the number of different configurations. By using resonators with high quality factor, we can generally consider windows with a reduced size and increase their number on a given frequency band. It is often instructive to divide the capacity of the tag by the frequency band used. This is the frequency density of the coding of chipless RFID tags. For example, when considering a frequency resolution of 50 MHz, it is possible to have a frequency density of around 10 bits per GHz.

The use of a large frequency band for coding is specific to chipless RFID. In fact, in traditional RFID, we saw that the most commonly used bands were ISM bands with specific standards (ISO) following the operating frequency. Chipless RFID, which does not use any communication protocols, is situated outside the ISO norms from the outset. That is why a simple and productive means of operating chipless RFID is to follow UWB standards. In these conditions, it is possible to cover a wide frequency band (several GHz) while using a short pulse signal. This way, it is possible to use chipless RFID tags that can contain upwards of 40 bits of information [VEN 12b].

1.3.2.3. Tag surface

In the example in Figure 1.11, increasing the number of resonators makes it possible to increase the coding capacity, but this necessarily translates into an increase in the tag's surface area. To limit (or take advantage of) couplings between the resonators, we must ensure that they are sufficiently (or correctly) spaced apart from each other. So, we can see that the quantity of information must be related to the geometric surface of the tags to be significant. This is the surface density of chipless RFID tags. In general, we estimate that to become a contender in the mass distribution market, chipless tags must be able to code 128 bits while retaining the dimensions of a credit card (85 mm × 55 mm). In terms of coding density, this translates into an objective of 2.73 bits per cm^2.

1.3.2.4. Read range

In the case of chipless RFID, the read range is determined by both the capacity of the tag to re-transmit enough power toward the reader and the sensitivity of the reader. If the RCS of the tag is known, we can calculate the theoretical read range thanks to the radar range equation. In practice, respecting the UWB standards, the read range is about 50 cm with a frequency approach [VEN 16]. In theory, distances of a few meters are possible.

1.3.2.5. Environmental sensitivity of tags

This final factor, although rarely taken into consideration, is very important. Chipless RFID shares its strong environmental sensitivity with traditional RFID. With traditional RFID, environmental disturbances have a tendency to deteriorate the tag's performance, although the information is not corrupted, thanks to both the durability of the temporal coding and the protective measures present in the communication protocol. In the case of chipless RFID, the absence of protocols and the use of frequency coding mean that a disturbance related to the environment can corrupt or even modify the information detected by the reader. In our example, we can see that for an environment that is very restricted, very large frequency windows should be included to guard against frequency shifts related to environmental disturbances. The increase in the width of the windows is necessarily accompanied by a decrease in the coding capacity. On the contrary, as a part of a tag designed to be very sensitive, this will allow the tag to be used a sensor instead of an identifier, thus adding an extra functionality to the tag.

1.3.3. *Positioning chipless RFID tags in relation to other automatic identification technologies*

The main properties of a chipless system are compared to barcodes and traditional RFID in the summary Table 1.4. Chipless RFID is a kind of compromise between these two technologies, combining certain functionalities of UHF RFID with the simplicity and very low cost of optical barcodes.

Chipless RFID shares its very low cost, totally passive label and printing-compatible manufacturing with barcodes. Because of this, they could be used in a similar way to optical barcodes, with labels that are directly printable by the user. The absence of a chip also means that all information contained in the tag is coded in a totally passive manner by the geometric form of the tag, which is similar to barcode coding.

Chipless RFID shares certain advantages related to radio-frequency communication with traditional RFID. The RF link makes it possible to reach a read range of around a meter with the tag in virtually any position, as well as the possibility of reading a tag without a direct line of sight (hidden object). The sensitivity of chipless tags to their environment makes it possible to use it as a sensor [KIM 13c]; this is also the case for traditional RFID. Contrary to traditional RFID, there is no possibility of using a communication protocol based on a precise synchronization of the clocks of the reader and the tag, and consequently it is difficult to do batch reading. Currently, a chipless tag can only be read because it does not contain any memory.

Family	Optical barcodes	Chipless RFID	Passive UHF RFID
Range (practical)	30 cm	< 1 m	10 m
Direct vision	Yes	No	No
Batch reading	No	No	Yes
Access	Reading	Reading	Reading/writing
Positioning	Any direction	Any direction	Any direction
Printable	Yes	Yes	No
Coding capacity	A few kbits	A few tens of bits	A few kbits
Sensor Functionality	No	Possible	Possible
Cost	> 0.005 euro	> 0.01 euro	> 0.2 euro

Table 1.4. *Positioning of chipless RFID in relation to other automatic identification technologies*

1.3.4. Conclusion and situating the study

The low quantity of information that it is currently possible to code in chipless RFID in relation to other identification technologies is an important observation. We estimate that the minimum quantity of information that chipless RFID must reach to compete with barcodes on the market of large-scale identification is 128 bits.

We will see in Chapter 2 that, for the moment, the coding methods proposed for chipless RFID do not allow us to reach that goal. We will also see that the quantity of "record" information registered for chipless tags has not evolved in recent years, which shows that traditional approaches have reached a kind of limit. To make it possible to pass from 50 bits (1.1×10^{15} different states) to the 128 bits necessary (3.4×10^{38} different states) it is not enough to optimize existing tags. New ways to code information on chipless RFID must be invented.

This book proposes a new approach that consists of coding the information on the entire frequency response of the tag and not only though distinctive elements like dips or peaks. The information contained in any signal is much richer than the presence of any element at a given frequency and this approach could make it possible to reach the desired 128 bits.

However, coding on the general form of the response requires designing tags whose RCS response is given in advance, which comes back to the inverse problem of the electromagnetic signature. Although benefitting from the simplified framework of chipless RFID (planar tags with small dimensions), this problem remains extremely difficult. In order to respond to this issue, the idea developed in this book is to construct an "objective" signature by assembling several more or less resonant patterns whose responses are known.

State of the Art of Chipless RFID Coding Methods

This chapter is an overview of the coding methods used in chipless RFID. The different lines of approach that have emerged around RFID are mentioned. A classification of chipless RFID tags based on coding method is proposed. The different types of coding are illustrated with examples from the literature. The objective of this chapter is to familiarize the reader with the physical principles in play for different types of coding and provide reference points in relation to performance.

2.1. Introduction

Traditional RFID has distinguished itself from barcodes in several fields including transportation, logistics and access control. Traditional RFID struggles to enter the large-scale distribution market due to the high cost of the tags, which is sometimes comparable to the price of the object it identifies. In order to reduce the unit cost of RFID tags, researchers have adopted a radical approach that consists of removing the silicon chip. This idea led to the development of an identification technology known as chipless RFID.

2.1.1. *Lines of research and positioning the study*

Chipless RFID is a recent area of research. The first articles mentioning this technology appeared in 2002. The first research projects demonstrated the possibility of coding information on a totally passive tag without an electronic chip and recuperating the data through RF interrogation.

The main obstacle for the development of this technology is the limited quantity of information that can be stored on a totally passive tag the size of a credit card. Since the initial projects, a large amount of research has been undertaken to address the problem of increasing the coding capacity and today there are many articles that treat this question. Currently, we can code up to around 50 bits, which is not yet sufficient as we estimate that 128 bits would be necessary for industrial applications. This remains the present challenge and the primary motivation for the development of amplitude coding and RCS synthesis presented in this study.

Over time, different lines of research have emerged around chipless RFID. Parallel to the increase in coding quantity, increasingly robust reading methods have been implemented to interrogate chipless tags[1]. There has also been research directed at developing low-cost readers adapted for reading chipless RFID tags [GAR 15, RAM 12, ZOM 15]. An important step that has been little explored until now is the transition to tags printed on low-cost materials like paper [VEN 13d, SHA 13, KIM 13c]. This step is difficult due to increased losses and lack of RF characterization with this type of substrate. Finally, we are seeing an increasing number of articles that address the use of chipless RFID tags as low-cost sensors[2].

The state of the art presented here focuses primarily on the question of how much information can be coded in a tag. Some articles related to other issues are referenced but will not be addressed in this chapter.

2.1.2. Classification of chipless RFID tags

From a technological perspective there is a large difference between RFID without chips and RFID with chips, and this must be taken into account when comparing the two approaches. The removal of the chip constitutes a substantial technological break with traditional RFID. The communication is no longer based on load commutation, so chipless RFID is a completely distinct technology. It also often uses different frequency bands

1 [VEN 13c, COS 15, KHA 15, PÖP 16b, REZ 14a, RAM 16b].
2 [VEN 15, KIM 13a, FEN 15, KIM 13b, COO 14, LAZ 16].

than traditional RFID. A chipless tag is more akin to a radar target where the information is registered in the geometric form of the tag. Sometimes the term "radio-frequency barcodes" is used in the literature.

Chipless RFID includes a wide variety of different tags that can seem to have little in common and are difficult to classify. However, we can distinguish two major families based on the techniques used for coding: tags whose information is coded in the temporal domain and those coded in the frequency domain.

Temporal coding is based on a principle of reflectometry. The tag is generally made up of a transmission line connected to an antenna. Parasitic elements are positioned along the line to create reflections at precise moments. The coding of the information is carried out by the presence or absence of a reflection at a given time. Temporal tags have a fairly large read range, in the order of a few meters, but they do not allow much information to be coded (less than 10 bits).

Frequency (or spectrum) coding is based on the use of resonant patterns that cause peaks (or dips) to appear in the spectrum of the reflected signal. The coding is achieved through the presence or absence of one of these distinctive elements at a given frequency. This type of coding requires a larger spectrum and mainly uses the UWB band (3–10 GHz), where regulations restrict the transmitting power to relatively low levels. Consequently, these tags have relatively low read ranges, in the order of 50 cm, but they do allow for coding greater quantities of information, in the order of about 50 bits.

Further on in this chapter, we will see that there are also "hybrid" types of coding that rely on several distinct physical quantities to code information. This coding almost always uses frequency coding. These types of coding appeared relatively recently in an attempt to increase coding capacity. Figure 2.1 charts the different types of coding used for chipless tags.

In the category of frequency tags, another feature that is fundamental for classification is the presence or absence of a ground plane in the tag's structure. The performances achieved by resonators with or without ground planes are very different and additional issues appear without a ground plane.

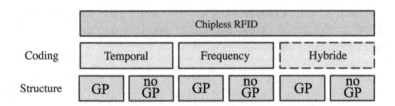

Figure 2.1. *Classification of chipless RFID tags.*
The main factors are the coding type and the presence or
absence of a ground plane in the tag's structure

Resonators with a ground plane resemble microstrip structures, which gives them a high RCS. The presence of a ground plane allows for good isolation from the object on which they are applied. These tags can be designed without prior knowledge of the objects that they will be applied to, which makes them "universal". However, it is very difficult to create these tags using a printing process on low-cost materials due to the presence of the two metallic layers and their high sensitivity to losses in the substrate. They are also much more difficult to hide. From an application standpoint, the issue of discretion is a very significant point that distinguishes chipless tags from barcodes which are necessarily visible.

In the absence of a ground plane, we generally observe less marked resonances which is synonymous with a weaker RCS. If this is the case, it is preferable to know the permittivity and the RCS level of the object that the tag will be applied to when it is being designed. However, it is much simpler to design a printed tag given its single layer structure.

The rest of this chapter presents a state of the art of different types of coding used for chipless RFID. It is not exhaustive on the topic because chipless RFID has been developing for 10 years and has a large number of publications. However, it will offer the reader a solid overview of the mechanisms and basic ideas used for coding chipless RFID. The reader can find supplementary information in other works dedicated to chipless RFID such as [PER 14, VEN 16, REZ 15a, PRE 12]. First, we will present some examples that will allow us to illustrate how temporal tags work. Next, we will examine frequency tags and the evolution that lead to their miniaturization. Today, as we mentioned, frequency tags have reached a plateau because we no longer see "records" appear in relation to coding capacity. In order to continue to increase the capacity of chipless tags, researchers have proposed an alternative approach whose effectiveness has yet to be demonstrated: hybrid coding. A few examples of tags designed using hybrid coding will be presented at the end of this chapter.

2.2. Tags coded in the temporal domain

The first chipless tags were directly inspired by traditional RFID where the coding is completed in the temporal domain. Researchers tried to reproduce temporal variations in the signal that appeared on the response frame of a traditional RFID tag. The chief difficulty for the design of a temporal chipless tag is to create transmission lines that are long enough to generate measurable delays while keeping the dimensions of the tag small. There is also the issue of losses along the transmission lines that reduce the amplitude of the reflections.

2.2.1. SAW tags

SAW (Surface Acoustic Wave) tags are probably the precursors of chipless RFID tags. Unlike the generations of chipless tags that followed, SAW tags were achieved using an expensive piezoelectric substrate that conflicted with the "low-cost" mentality of chipless RFID. However, they were still the first tags without a chip that appeared in the literature [HAR 02] and they illustrated the principle of coding in the temporal domain. SAW tags are the only chipless RFID tags commercialized to date.

SAW tags generally operate at 2.45 GHz [HAR 02, PLE 10]. The coding capacity achieved by this type of tag is 256 bits, which is compatible with the EPC standard and comparable to traditional RFID. With SAW tags, the reader emits a pulse with a power of 10 mW in line with ETSI regulations. With traditional RFID, it is necessary to provide the chip with enough power to activate and operate it. At an equal read range, the SAW tag requires a much lower power than traditional RFID. This type of tag can reach a read range in the order of a few meters. However, the cost of the tag is very high because of the piezoelectric properties of the substrate. These tags are not printable on low-cost materials like paper.

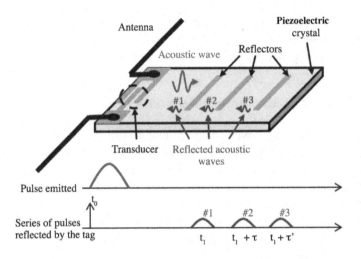

Figure 2.2. *Operating principle of an SAW tag presented in [PLE 10]. For a color version of this figure, see www.iste.co.uk/rance/rfid.zip*

Figure 2.2 illustrates the operating principle of an SAW tag. The incident electric wave is converted into a surface acoustic wave through a transducer connector interface (TCI). The surface acoustic wave propagates along the piezoelectric crystal and is reflected by different reflectors, which creates a series of pulses separated by a certain time delay. The series of pulses is reconverted into electromagnetic waves through the TCI and re-transmitted toward the reader where the identifier of the tag is decoded. The coding is achieved by modifying the position of the reflectors on the substrate, which has the effect of modifying the time delay between the reflected pulses. This type of coding is sometimes called PPM (pulse position modulation), referencing signal theory.

The SAW tag uses the unique nature of the piezoelectric material to transform the electromagnetic wave into an acoustic wave that is 10^5 times slower. SAW tags act as delay lines that provide an easily measurable delay while preserving relatively small dimensions (10 mm × 10 mm, not counting the antennas).

In the case of a low-cost non-piezoelectric substrate, a longer transmission line is necessary to produce a measurable delay. It is therefore more difficult to obtain significant coding densities as we will see in the examples in the next paragraph.

2.2.2. *Transmission line tags*

Several attempts have been made to adapt SAW tag coding principles to less costly substrates. The coding method remains similar but the amount of information coded by the tag is very limited.

The tag created by Zheng *et al.* [ZHA 06] in 2006 is a classic example. The tag and its equivalent electrical circuit are represented in Figure 2.3. This tag has a maximum coding capacity of 8 bits for an area of 8.2 cm × 10.6 cm. Unlike the SAW tag that operates at 2.45 GHz, this tag uses a UWB frequency range and the interrogation signal is a pulse with a duration of 2 ns. The design relies on several sections of micro-strip transmission lines on Rogers 4350. Each section of transmission line codes one bit and requires a minimum length of 18 cm to avoid temporal overlapping of the reflected pulses. The lines are arranged in the form of meanders to minimize the total surface of the tag. The lumped [ZHA 06] or distributed [ZHE 08] capacitors are positioned between each section of line in order to create the impedance mismatches that are at the origin of the reflections.

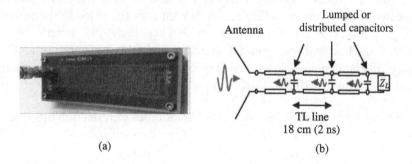

(a)   (b)

Figure 2.3. *Temporal coded tag created by Zengh et al. [ZHE 08];*
a) Delay line in meander (without antennas); b) Equivalent electrical circuit.
For a color version of this figure, see www.iste.co.uk/rance/rfid.zip

The authors showed that they were able to configure the tags through the localized addition of material using a conductive ink jet printer. However, this design is still too large for tracking applications on consumer products.

Even when considering a tag with larger dimensions, it is difficult to extend the transmission line network due to losses related to propagation. This type of approach is therefore limited to low coding capacities. The theoretical limits of temporal tags based on the principle of reflectometry were recently assessed in [PÖP 16a].

2.2.3. *Variable terminal impedance tags*

Given the inherent limitation to the number of bits coded by a temporal tag, other types of applications have been considered such as localization [HU 10] or the production of low-cost passive sensors [GIR 12b, SHR 09]. The tags presented in these publications also use the temporal domain to code information but unlike the tags presented above these tags only have one segment of transmission line, making it possible to limit losses related to propagation. The coding is done by modifying the terminal impedance of the line which modifies the form of the reflected pulse. This is the basic principle of coding used in traditional RFID where the chip varies between two different impedance states. Variable terminal impedance tags can be either narrowband [SHR 09] or wideband [RAM 11].

The tag achieved in [HU 10] is represented in Figure 2.4 as well as its equivalent electrical diagram. This tag is made up of a wideband antenna connected to a transmission line that ends with a localized load. Six different configurations can be created by modifying both the length of the transmission line and the terminal impedance condition according to three types of loads: matched, short-circuit or open-circuit. The temporal response of the tag for these different loads is represented in Figure 2.5.

This tag's operation relies on two different types of reflection mechanism that are represented in Figure 2.4(b). When an incident pulse encounters the tag, a part of the power is immediately reflected by the structure of the antenna according to a specular reflection: this is the structural mode of the tag. The corresponding temporal response to this mode is an image of the incident pulse and does not depend on the value of the load impedance. In Figure 2.5, the part of the response related to the structural mode appears around the time $T_1 = 5.328\ ns$. A second mechanism appears because part of the power is effectively collected by the antenna and then guided toward the load through the transmission line. Part of this power is then reflected by the load: this is the antenna mode of the tag. The response associated with the antenna mode depends on both the length of the line and the value of the terminal impedance. It corresponds to the part of the responses centered around $T_2 = 5.843\ ns$ in Figure 2.5.

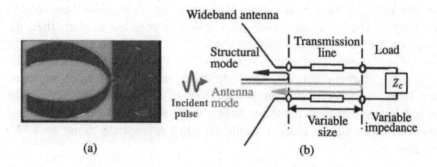

(a)   (b)

Figure 2.4. *a) Structure of variable terminal impedance tags created in [HU 10]; b) Equivalent electrical diagram. For a color version of this figure, see www.iste.co.uk/rance/rfid.zip*

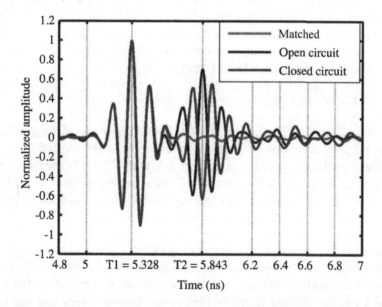

Figure 2.5. *Temporal response of the tag [Hu 10] for different terminal impedance conditions. The structural mode corresponds to the part centered around T1 (the part of the signal that is identical regardless of the load) and the antenna mode to the part centered around T2. For a color version of this figure, see www.iste.co.uk/rance/rfid.zip*

In the absence of several segments of transmission lines, the structural mode is traditionally used as a reference element to calculate the delay related to the length of the line and, in so doing, completes a PPM coding. The transmission line must induce a sufficiently large delay (at least the duration of a pulse) to make it possible to clearly distinguish between the two modes. The structural mode also makes it possible to normalize the power of the reflected signal (see Figure 2.5) in order to eliminate the dependence (attenuation) of the response with respect to the read range. Still, to increase the number of states different impedance values must be considered. In order to remain in the spirit of chipless RFID, the use of lumped loads (like here for the matched load) should not be used for cost reasons. In fact, in this case, it would be more appropriate to use an RFID chip directly.

Although they hold a limited quantity of information, temporal terminal impedance tags lend themselves to the creation of communicative passive sensors. For example, in [SHR 09], the authors show that it is possible to add a remote communication functionality to an ethylene sensor simply by connecting it to the tag's termination. Other studies [GIR 12b] have shown that the tag itself can serve as a sensor if using a substrate with a permittivity that is sensitive to an environmental parameter. A detailed study of this kind of application can be found in [RAM 16a].

2.3. Tags coded in the frequency domain

The second approach relies on the frequency signature of a tag to code information. The frequency signature corresponds to the evolution of the amplitude or to the phase of the tag's response with respect to the frequency. The coding is based on the positioning of distinct elements like the peaks or dips of the spectral signature. Frequency coding uses resonant elements whose frequency is adjustable with a geometric parameter.

Researchers agree that frequency coding offers a greater coding capacity and density than the temporal approach. Frequency tags operate on the UWB band and are therefore subject to radiation standards imposed by the FCC (Federal Communications Commission) in the United States and the ECC (Electronic Communications Committee) in Europe. Consequently, frequency tags generally have weaker read ranges (≤ 1 m) than temporal tags. The fact that they can hold a large amount of information makes them more sensitive to the environment and extracting the identifier often requires a calibration step when measuring.

Frequency tags have evolved a great deal since they appeared in 2008. We note three consecutive steps that are represented in a diagram in Figure 2.6 that enabled the miniaturization of the tags and consequently allowed them to reach ever-higher coding densities.

The first types of frequency tags are based on a guided circuit approach (Figure 2.6(a)). They have a receiving antenna and a transmitting antenna connected by a planar filter that ensures the coding of the information. The tags achieved with this type of approach are fairly bulky due to the two antennas. In order to reduce the surface area, a second approach (Figure 2.6(b)) has been proposed that consists of using the same antenna for transmitting and receiving. The antenna is generally loaded by resonant elements that ensure the filter's operation. The most recent approach that provides the best coding densities uses resonant scatterers that fulfill the functions of a receiving antenna, filter and transmitting antenna by themselves (Figure 2.6(c)). This means that there is no longer an antenna in the structure strictly speaking. These resonant scatterers are referred to as Radiofrequency Encoding Particles (REP). This approach is what will be used in the rest of this study [PER 14, VEN 16].

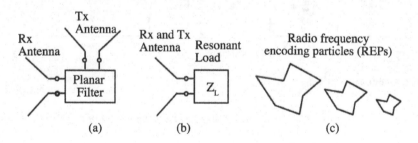

Figure 2.6. *Evolution of frequency tags; a) Planar filter tags; b) Single antenna loaded with resonant elements; c) Radiofrequency encoding particles (REPs)*

2.3.1. *Planar filter tags*

This approach is the first to have been studied in detail in the development of frequency chipless tags. It demonstrated that frequency tags make it possible to reach high quantities of information almost comparable to that of EAN 13 optical barcodes. It has encouraged researchers to explore the possibilities of frequency coded tags.

The first appearance of this concept was in 2008 in [PRE 08]. The tag proposed (Figure 2.7(a)) is composed of two wideband antennas linked by a microstrip line. Spiral resonators of different lengths are positioned along the line and create dips in the tag's spectral response (Figure 2.7(b)). The coding is achieved by the presence or absence of a dip at a given frequency in the spectrum. Consequently, there is a 1:1 correspondence between the number of resonators and the number of bits coded by the tag. Rather than removing the resonators to eliminate the dips, the authors demonstrate that it is possible to short-circuit them to move the resonance to higher frequencies outside the operating band. This idea makes it possible to consider tags that can be configured using conductive ink printing and to limit the impact of couplings between resonators for each new configuration. The transmitting and receiving antennas are in cross-polarization in order to limit the couplings between the antennas. This also contributes to improving the robustness of the reading in relation to its environment.

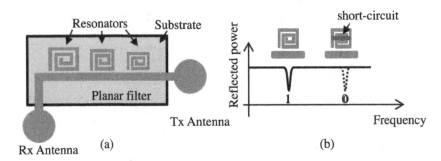

Figure 2.7. *Structure; a) and coding principle: b) of a frequency tag with two antennas [PRE 08]*

Although the initial tag had a coding capacity limited to 6 bits, the same structure was optimized in 2009 [PRE 09a] to reach a capacity of 35 bits on an 88 mm × 65 mm surface. This tag operates on a frequency band of 3 to 7 GHz, which is within the UWB band. Compared to the previous tag, the transmission line is arranged in a meander form and the resonators are positioned on either side of the line (Figure 2.8), which makes it possible to significantly reduce the area occupied by the tag.

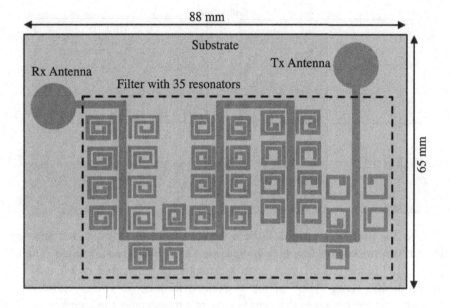

Figure 2.8. *A 35-bit tag structure presented in [PRE 09a]*

The two tags presented above use a microstrip line and are therefore hardly compatible with a printing realization process. In order to make this structure potentially printable, the authors proposed a variant based on a coplanar waveguide (CPW) transmission line in a flexible substrate [PRE 09b]. The tag created in that way achieved a capacity of 23 bits on a 108 mm × 64 mm surface.

The tag presented in [PRE 09a] has been reused in [PRE 09c] to show that it is possible to achieve coding on the phase of the frequency response. In this study, the authors demonstrate experimentally that information transmitted by phase is more robust to noise than that transmitted by amplitude, which makes it possible to reach greater read ranges.

Based on this observation, several studies were carried out on tag systems based on group delay, including [NIJ 12, NAI 11, GUP 11]. The structure of the tags proposed in [NIJ 12, NAI 11] is represented in Figure 2.9.

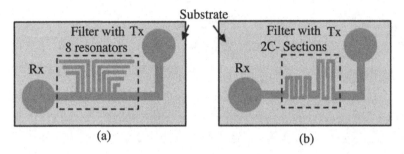

Figures 2.9. *Planar filter tags whose coding was achieved using group delay; a) Structure of the 8-bit tag presented in [NIJ 12]; b) Structure of the 2-bit tag presented in [NAI 11]*

Like the previous examples, these tags have two cross-polarized antennas linked by a microstrip line on which resonant elements are arranged. The tag [NIJ 12] (Figure 2.9(a)) uses open quarter-wave stubs that cause peaks to appear in the group delay of the resonance frequency. There is also a 1:1 correspondence between the number of resonators and the number of bits coded. However, this tag offers a more robust reading than its predecessor: it has more distinct dips (6 dB for amplitude) and a certain form of redundancy because the information can also be decoded using group delay. In the case of the tag in [NAI 11], the C-sections cause peaks to appear in the group delay for certain frequencies. By modifying the size of the C-sections, it is possible to cause the amplitude of the peaks to vary, which translates into a more or less delayed signal at the temporal level. This tag is also less sensitive to additive noise.

2.3.2. *Tags using a loaded wideband antenna*

The presence of two different cross-polarized antennas for receiving and transmitting has the advantage of limiting interference between the signal received and the response of the tag. However, this solution has the disadvantage of taking up a larger footprint than a single antenna. It also limits the maximum number of resonators due to the insertion losses in the filter. There is another approach in which a single wideband antenna is used for both receiving and transmitting. The resonant elements can therefore be directly incorporated into the structure of the antenna which allows for the removal of the transmission line.

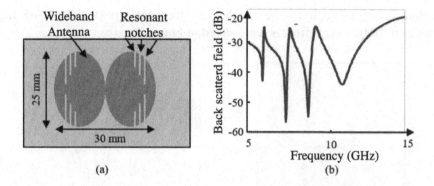

Figure 2.10. *Tag with a single dipole antenna loaded with resonant notches [BLIS 11]: a) Geometric structure of the tag; b) Frequency response of the tag*

An example of this type of tag was proposed in 2009 and explained in detail in an article in 2011 [BLIS 11]. This tag operates on the UWB band and does not have a ground plane. The tag is made up of an elliptical dipole in which there are notches (Figure 2.10(a)). The isolated dipole presents a wideband signature. When the notches are prepared, the dipole antenna is loaded with these resonant elements which then act as a band rejector filter directly integrated in the antenna. The electromagnetic signature of the tag presents very selective dips at the resonance frequencies of the notches (quarter-wave resonators). In the example presented, three notches with different lengths are used, repeating four times in order to increase the depth of the dips. The coding capacity of the tag is 3 bits for an area of 25 mm × 30 mm. This article also presents a clear theoretical interest because it uses the SEM (Singularity Expansion Method) theory for the design and detection of the tag's identifier. The principle of SEM will be presented in the next chapter and illustrated with an example in the framework of chipless RFID (section 3.5.2).

A similar tag optimized for coding capacity was proposed in 2014 [REZ 14b]. Like in the previous example, the tag does not have a ground plane and operates on the UWB band. The tag is made up of a circular wideband antenna on which 24 notches of different sizes are integrated (Figure 2.11(a)). The behavior of the notches has been studied theoretically [REZ 15b]. Each notch behaves like a quarter-wave resonator and introduces a dip in the tag's signature. In order to modify the tag's identifier, the authors demonstrate that it is possible to eliminate anti-resonance simply by adding a short-circuit to the notch. The frequency response of the tag obtained for two identifiers is represented in Figure 2.11(b).

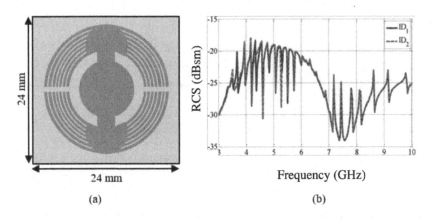

(a)   (b)

Figure 2.11. *A 24-bit tag presented in [REZ 14b]; a) Structure of the tag; b) Frequency response of the tag for two different identifiers. For a color version of this figure, see www.iste.co.uk/rance/rfid.zip*

The tag studied has a coding capacity of 24 bits on an area of 24 mm × 24 mm, which gives a coding density of 4.1 bits per cm², well above that obtained by the best planar filter tags presented in the previous section.

2.3.3. *RF encoding particle approach*

The final step to minimize the area of chipless tags is to combine the functions of the receiving antenna, transmitting antenna and filter into a single element that we can call a filtering antenna or simply a resonator.

In 2005, a very simple solution was proposed in [JAL 05], which includes a ground plane. Microstrip dipole resonators of varying lengths are used to transcribe the principle of barcodes into RF, as illustrated in Figure 2.12(a). Given the presence of the ground plane, these dipoles act as patch antennas or short-circuit microstrip dipoles which gives them a significant RCS and good selectivity. These microstrip dipoles act like half-wave resonators. Each dipole is a resonant antenna whose response is in phase opposition to that of the ground plane (wideband), which results in dips in the overall response of the tag (Figure 2.12(b)). The dipoles no longer need to be connected to a wideband antenna structure, which makes the structure easily extendable. The same structure has been used by adding variable capacitors to the microstrip dipoles. The link that exists between the value of the capacitor and the resonance frequency makes it possible to code information.

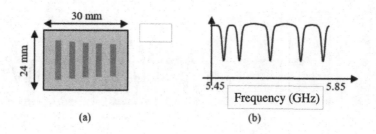

(a)   (b)

Figure 2.12. *Basic example of a tag realized with the RF encoding particle method; a) Structure of the tag; b) Frequency response of the tag*

The coding capacity reached by the structure proposed by [JAL 05] is in the order of 5 bits using five resonators on a slightly smaller surface of around 25×30 mm^2. The necessary frequency range is between 5.45 GHz and 5.85 GHz, or 1 bit per 100 MHz. Measurements carried out show that it is possible to detect this tag at a distance of up to several tens of cm for a transmitting power of 500 mW.

The RF encoding particle (REP) approach was introduced in [VEN 11]. The method consists of viewing the tag as a radar target that is designed to present resonances at given frequencies. This breaks away from the idea of an antenna and a transmission line to guide and then filter the signal as desired.

Another example of a tag designed using the same principle was proposed in 2012 in [VEN 12b]. This tag operates between 2 and 4 GHz, has no ground plane and is realized on FR4. The RF encoding particles are C-shaped resonators. The C-resonator has been selected from among other configurations in [VEN 11] because it presents a good compromise in terms of RCS levels, miniaturization and selectivity. The C-resonators act as quarter-wave resonators and the resonance frequency can be controlled by modifying the length of their arms. The tag proposed has 20 resonators stacked vertically as represented in Figure 2.13(a). To configure the tag, the notches can be short-circuited, which rejects the resonance frequency beyond the operating band. This approach makes it possible to keep couplings relatively similar from one configuration to another. The response of the tag for three different identifiers is represented in Figure 2.13(b).

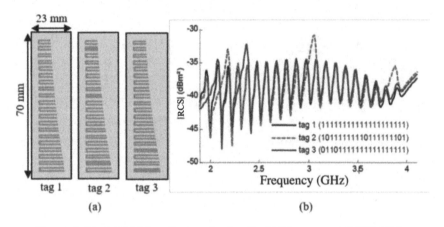

(a)   (b)

Figure 2.13. *A 20-bit tag designed using the REP approach [VEN 12b];*
a) Structure of the tag configured for three different identifiers;
b) Response of the tag to these three identifiers. For a color version of this figure,
see www.iste.co.uk/rance/rfid.zip

Like in the previous case, there is a 1:1 correspondence between the number of resonators and the number of bits coded. The tag reaches a capacity of 20 bits for an area of 25 mm × 70 mm which gives it a coding density of 1.14 bits per cm². The tag is easily extendible but the frequency band used is limited by the appearance of resonances due to higher modes starting from 6 GHz.

The article also addresses an important point concerning tags that are realized without ground planes. Without a ground plane, the tag is very sensitive to the object it is applied to and significant frequency shifts in the response of the resonators can be observed. If the tag is not designed knowing the permittivity of the object in advance, this can create detection errors. A method of compensating for the frequency shifts is proposed in [VEN 12b]. The method is to use resonators at both ends of the frequency band like sensors in order to evaluate the permittivity of the object.

Another tag created using the RF encoding particle approach was presented in 2012 in [VEN 12c] and discussed in detail in [VEN 12d]. This tag has a ground plane, operates on the UWB band and is realized on Roger RO4003. The encoding particles used for the coding are circular patch resonators that overlap each other in order to optimize the area occupied (Figure 2.14(a)). This type of structure resonates when the wavelength is equal to one quarter of the circle's perimeter. It is therefore possible to adjust the resonance frequency simply by modifying the value of the radius. The presence of a ground plane allows for good isolation between the tag and the object and is also responsible for the substantial selectivity of the resonators. The response of the tag is represented in terms of amplitude (Figure 2.14(b)) and group delay (Figure 2.14(c)). Near the resonance frequency the response presents a dip, revealing a destructive interference between the structural mode and the antenna mode. The 12 resonances can be identified using amplitude or group delay.

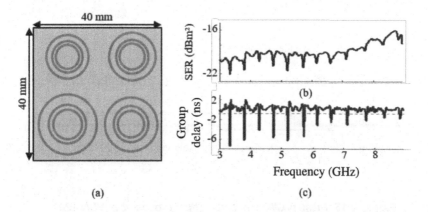

Figure 2.14. *Tag that is insensitive to the polarization presented in [VEN 12c]; a) Structure of the tag; b) Response of the tag in terms of amplitude; c) Response of the tag in terms of group delay*

Given the rotational symmetry of this type of resonator, the tag does not depend on the polarization of the incident wave. It is, to our knowledge, the only case of a chipless RFID tag independent of polarization published in the literature.

For the tags presented so far, coding was completed with the presence or absence of a peak at a predefined frequency. Consequently, there is a 1:1 correspondence between the number of bits and the number of resonators present in the tag. In [VEN 12d], the authors propose a different approach that makes it possible to considerably increase the quantity of information coded by the tag. The method consists of attributing a certain frequency range, divided into frequency slots, to each resonator (see Figure 2.15). Each resonator can then code different states corresponding to the number of available slots. In the case of the tag presented in [VEN 12c], the total frequency range is the UWB band (3.1 GHz to 10.6 GHz) and the authors estimate that a frequency resolution of 30 MHz is sufficient to distinguish two frequency slots. Consequently, there are 250 frequency slots to be shared between the 12 resonators present on the tag. The authors also provide for three isolation slots between each resonator to avoid any physical overlay. Each resonator can therefore use 17 slots for coding which gives a capacity of 17^12 different identifiers. This approach makes it possible to reach a record coding capacity of 49 bits for a tag with dimensions of 40 mm × 40 mm.

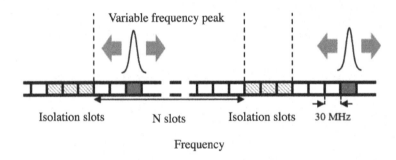

Figure 2.15. *Pulse position coding method proposed in [VEN 12d]*

2.4. Hybrid tags

One possibility for increasing coding capacity is to use an RF encoding particle capable of coding different information through two distinct physical quantities. This approach seems promising because it makes it possible to break away from the linear dependence that exists between the number of resonators and the number of bits coded. In the case where two physical quantities are used a quadratic relation seems to appear. However, resonators designed in this way are often more complicated and larger than when a single physical quantity is used. Although this approach generally provides a greater number of states coded at the level of an encoding particle, this does not necessarily mean that the surface density of the coding increases. The hybrid coding method, although promising, must still be demonstrated because, for the moment, no hybrid tags have succeeded in surpassing the 49 bits obtained in the work of Vena *et al.* [VEN 12c]. Because frequency coding is both easy to implement and performs well in terms of coding capacity, it has been associated with a second physical quantity.

Hybrid coding made its first appearance in the literature in 2012 in [VEN 11]. The tag presented does not have a ground plane and operates on the 2.5–7.5 GHz band and is realized on a FR4 substrate. The encoding particle used is the C-shaped resonator that was already described in [VEN 12b] and which is represented in Figure 2.16(a).

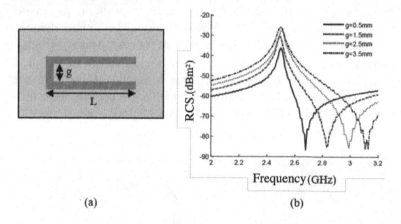

(a)   (b)

Figure 2.16. *a) Structure of the C-shaped resonator; b) Frequency response of the C-shaped resonator for different values of the parameter g. For a color version of this figure, see www.iste.co.uk/rance/rfid.zip*

The C-resonator is particular because the resonance peak is followed by a dip in its signature (Figure 2.16(b)). The frequency of the peak is mainly controlled by the length of the metallic wires (L in Figure 2.16(a)) and the difference between the peak and the dip depends on the size of the slot (g in Figure 2.16(a)). Therefore, it appears possible to control these two quantities from those two geometric and independent parameters. As an example, the signature of an isolated C-resonator for four different gap values is represented in Figure 2.16(b)). The final tag has five C-resonators stacked vertically for an area of 20 mm × 40 mm. With this approach, each resonator is capable of coding six states linked to the position of the peak and four

states linked to the difference between the peak and the dip, for a total of
$6 \times 4 = 24$ states. The overall tag has a coding capacity of 22.9 bits for a
surface coding density of 2.9 bits per cm².

An example of a tag with hybrid coding using both the frequency and
amplitude of a peak was published in 2013 in [VEN 13a]. The objective of
the article is to create a tag whose configuration cannot be seen by the naked
eye in order to respond to issues in the fight against counterfeit goods. The
coding aspect, although it is discussed, is not the focus of the article and the
authors do not particularly try to optimize the tag from a coding capacity
standpoint. In this article, the tag is created using inkjet printing on a flexible
polyamide substrate. Loop resonators in figure eights are used as encoding
particles (Figure 2.17(a)). The resonance frequency of the loops is modified
by applying a change of scale to the base structure. The authors demonstrate
that it is possible to modify the amplitude of the peaks by adding a
transparent resistive band of variable width to the center of the structure
(Figure 2.17(b)). The overall tag has three resonators for a total area of
70 mm × 40 mm. Each resonator is capable of coding four different
amplitude states which makes it possible to reach a total coding capacity of 6
bits for only three resonators.

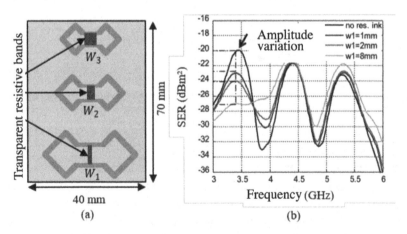

Figure 2.17. *a) Structure of the hybrid frequency-amplitude tag proposed in
[VEN 13a]; b) spectral response of the tag for different values of W1. For a color
version of this figure, see www.iste.co.uk/rance/rfid.zip*

Another tag using hybrid coding was proposed in the literature in 2015 [ELA 15]. The authors propose to code information at the level of both the resonance frequency and the quality factor. To do that, the tag incorporates different kinds of resonators. Although the idea is interesting, the research presented is not very successful. The method of control is based on an empirical approach that seems computationally intensive and the authors have a difficult time assessing the coding capacity obtained.

Although it does not objectively belong to the hybrid coding tags category, it seems relevant to mention certain articles that adopt a similar philosophy. This means that, like hybrid coding, the basic idea of the design is to code a large number of states with a single resonator, which is created by adjusting the value of several independent geometric parameters.

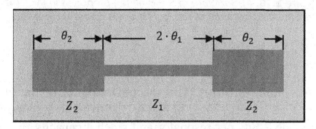

Figure 2.18. *Structure and geometric parameters of a stepped impedance resonator*

An example of this type of tag was proposed in [NIJ 14]. The tag operates on the UWB band and has a ground plane and a duroid RT substrate. The encoding particle used is a stepped impedance resonator like the one represented in Figure 2.18. The particularity of this type of resonator is that it is possible to independently control the fundamental and the first harmonic resonance frequencies by acting on both the aspect ratio (θ_1, θ_2 in Figure 2.18) and the impedance ratio (Z_1, Z_2 in Figure 2.18) between the different sections of the line [CHE 06]. This way, a single resonator can be used to cause two peaks to appear in the spectral response of the tag, making it possible to consider interesting surface coding densities.

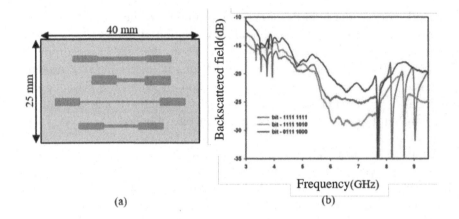

Figure 2.19. *a) Structure of the 8-bit tag proposed in [NIJ 14];*
b) Spectral response of the tag for three different identifiers.
For a color version of this figure, see www.iste.co.uk/rance/rfid.zip

The overall tag proposed in [NIJ 14] has four stepped impedance resonators (see Figure 2.19(a)). The spectral response of the tag for three different identifiers is represented in Figure 2.19(b). The tag proposed makes it possible to code 2 bits per resonator and reach a capacity of 8 bits for an area of 40 mm × 25 mm. It has a surface coding density of 1.06 bits per cm². Using a pulse position coding like the one presented in [VEN 12d], the authors estimate that it is possible to reach a capacity of 39 bits using only four resonators.

Another tag with a relatively similar philosophy to hybrid coding was presented in [ISL 15]. Here, the authors use a tag with different information following the polarization (vertical or horizontal) of the incident wave. The tag proposed does not have a ground plane and operates on the 6–12 GHz band. It is based on a square wideband antenna inside of which there are vertical and horizontal notches of varying lengths (Figure 2.20(a)). The response of a vertical slot is the maximum for an incident wave polarized vertically and zero if the incident wave is polarized horizontally. According to this principle, it is possible to separately detect information contained in the vertical and horizontal notches of the tag, as represented in Figure 2.20.

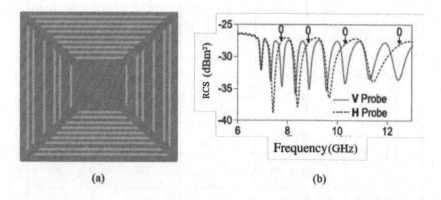

Figure 2.20. *a) Structure of the dual-polarized tag presented in [ISL 15]; b) Spectral response of the tag according to the vertical or horizontal polarization. The tag presented corresponds to the identifier: V-111111111 H-110101010. For a color version of this figure, see www.iste.co.uk/rance/rfid.zip*

This approach makes it possible to improve the spectral coding density for a given frequency band but it requires using antennas with polarization agility. The capacity reached by this tag was 18 bits for an area of 21 mm × 21 mm.

2.5. Conclusion

The properties of tags with the highest coding capacities are summarized in Table 2.1. This table shows several things. With the exception of SAW tags, it appears clear that frequency coded tags make it possible to reach coding capacities that are much greater than temporally coded tags. The radiofrequency encoding particle method (REP in the table) seems to be the most promising in terms of coding capacities, especially when it is paired with pulse position coding like the one presented in [VEN 12d]. The hybrid coding principle makes it possible to code more bits per resonator but not necessarily to reach higher surface coding densities than direct frequency coding. The relevance of this approach still needs to be demonstrated. Tags without a ground plane present lower coding capacities and weaker read ranges than tags with a ground plane.

	Type of tag	Ref	Coding capacity (bits)	Surface (mm²)	Bits per resonator	BP	Range	Ground plane
Temporal	SAW Tag	[PLE 10]	256	10 × 10 + ant	-	2.45 GHz	30 m	No
	Delay line	[ZHE 08]	8	82 × 106 + ant	-	UWB	NR	No
	Variable impedance line	[HU 10]	2.5	23 × 23	-	UWB	NR	No
Frequency	Microstrip planar filter	[PRE 09a]	35	88 × 65	1	UWB	1 m	Yes
	CPW planar strip	[PRE 09b]	23	108 × 64	1	UWB	NR	No
	Antenna loaded with notches	[REZ 15a]	24	24 × 24	1	UWB	NR	No
	REP: C-resonators	[VEN 12b]	20	25 × 70	1	2–4 GHz	50 cm	No
	REP: Circular microstrip resonator	[VEN 12c]	49	40 × 40	4.1	UWB	1 m	Yes
Hybrid	REP: C-resonators	[VEN 11]	21.9	20 × 40	4.4	2.5–7.5 GHz	50 cm	No
	REP: stepped impedance resonator	[NIJ 14]	39	40 × 25	9.8	UWB	NR	Yes
	Dual-polarized	[ISL 15]	18	21 × 21	1	6–12 GHz	NR	No

Table 2.1. *Comparison of the coding capacity of tags presented in this chapter according to their coding type*

Theory of Chipless RFID Tags

This chapter includes a certain number of theoretical concepts that apply to chipless RFID. These concepts come from several fields, such as radar, antenna, and traditional RFID. First, we provide a precise definition of what the "response of a chipless tag" is and then we present the reading system that makes it possible to measure them. After establishing the physical definition of the response, we examine its different components. The component that is useful for coding is based on the resonant nature of the constituent parts of the tag, so we specify the characteristic quantities that make it possible to analyze these resonances. Finally, we demonstrate that the resonant nature of the response enables the implementation of techniques to improve the readability of a chipless tag. This chapter presents a general perspective of the physical phenomena involved in interrogating a chipless tag. The concepts addressed here will be used in the following chapters concerning the design of chipless RFID tags.

3.1. Response of a chipless RFID tag

The response of a chipless tag can be described using different physical quantities depending on the type of coding used. The Radar Cross Section (RCS) is generally used when information is only provided by the magnitude of the response. RCS is a scalar quantity, which means that it does not contain any information about the phase of the tag's response. It also generally depends on the polarization of the transmitting and receiving antennas. When the information is linked to either the polarization or the phase of the reflected wave, we generally choose to describe the answer by its polarimetric scattering matrix. These two quantities, especially the second one, although traditionally used in the domain of radar, are not very common in the field of antennas, so we will revisit their definitions.

3.1.1. *Radar Cross Section (RCS)*

When a chipless tag is placed in the beam of a reader, it scatters some of the incident energy in all directions. This phenomenon is called *scattering*. The spatial distribution of the field that results depends on the size, shape and composition of the tag as well as the arrival direction and the nature of the incident wave. From a physical standpoint, a tag is therefore a *scatterer*. Generally, the direction of the incident wave is different from the direction of observation. This is called *bistatic* scattering. Concretely, these two directions are often identical because the transmitting antenna and the receiving antenna of the reader are placed very close to one another or even overlap. This is called *monostatic* scattering or *backscattering*. These definitions are illustrated in a diagram in Figure 3.1.

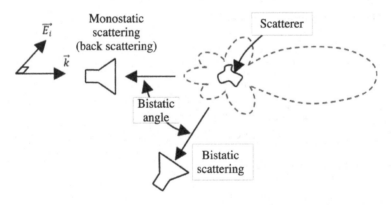

Figure 3.1. *Monostatic and bistatic scattering. The antennas represented are used for receiving. The direction and the polarization of the incident wave are represented by the vectors $\vec{k}$ and $\vec{E}_i$ respectively*

Observing the distribution of fields around the tag is a complex task. In the framework of radar, it is often preferable to describe the characteristics of the echo using a fictitious surface: a Radar Cross Section (RCS) [KNO 04]. RCS is a fictitious surface that often has no relation to the physical surface of the tag.

By definition, the RCS is a far-field parameter that relates two power densities, one of which is measured at the tag, while the other is measured at the receiving antenna, as in Figure 3.2. We assume that the reader is sufficiently distant from the tag and that the tag is small enough for the incident wave and scattered wave to be spherical. The tag exposed to a wave behaves like an antenna and captures part of the power of the wave through an effective capture surface, σ. The power captured, P, is calculated as the product of the power density at the tag, W^i, with the capture surface:

$$P = \sigma W^i \qquad [3.1]$$

For the definition of the RCS, it is assumed that the tag radiates all of the captured power isotropically. If the target is small compared to the observation distance R, the scattered power density, W^S, decreases with R following:

$$W^s = \frac{P}{4\pi R^2} = \frac{\sigma W^i}{4\pi R^2} \qquad [3.2]$$

Although the backscattered wave is measured at a great distance from the tag, no specific direction has been imposed. We now express σ based on other quantities:

$$\sigma = 4\pi R^2 \frac{W^s}{W^i} \qquad [3.3]$$

The capture surface σ defined in this way is the RCS of the tag. In practice the tag is rarely isotropic, which means that the RCS varies according to the incident wave direction and the observation direction.

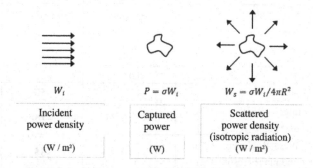

W_i	$P = \sigma W_i$	$W_s = \sigma W_i/4\pi R^2$
Incident power density	Captured power	Scattered power density (isotropic radiation)
(W / m²)	(W)	(W / m²)

Figure 3.2. *Power densities involved in the definition of the Radar Cross Section*

The power densities of the incident wave and the scattered wave can be expressed as the products of electric and magnetic fields:

$$W^i = E^i \cdot H^i/2 = Y_0 \cdot |E^i|^2/2$$

$$W^s = E^s \cdot H^s/2 = Y_0 \cdot |E^s|^2/2 \qquad\qquad [3.4]$$

where E^i and H^i are the intensity of the incident electrical and magnetic fields, E^s and H^s are the intensity of the scattered electric and magnetic fields, and Y_0 is the characteristic admittance of free space. If E and H are measured in volts per meter and amperes per meter respectively, the power density is expressed in watts per square meter. By introducing the expression of W^i and W^S in [3.3], we can express σ in relation to the fields:

$$\sigma = 4\pi R^2 \frac{|E^s|^2}{|E^i|^2} \qquad\qquad [3.5]$$

The dependence on R in the expression [3.5] is only superficial. Because it is in the far-field area, the intensity of the scattered field decreases in an inversely proportional way to R, which implicitly results in a term in R^2 at the denominator that offsets the term in R^2 of the numerator. In practice, the far-field condition must be respected to measure the RCS of a chipless tag.

The typical RCS values obtained for different classes of radar targets [SKO 08] are presented in Table 3.1. These values are indicative. Within the same class, we can expect to find RCS variations up to about 20 decibels depending on the frequency, incidence angle, or specific properties of the target.

Target	RCS (m²)	RCS (dBsm)
Warship	$5 \cdot 10^3$	37
Plane	10^2	20
Person	10^0	0
Bird	10^{-2}	-20
Insect	10^{-5}	-50
Chipless RFID tag with ground plane (resonance)	$3 \cdot 10^{-2}$	-15
Chipless RFID tag without ground plane (resonance)	$3 \cdot 10^{-4}$	-35

Table 3.1. *Indicative values of the RCS of different categories of radar targets*

3.1.2. *Polarimetric scattering matrix*

The RCS as a scalar quantity does not provide all of the information about the reflecting power of an object. In the early 1960s, the relative phase and the polarimetric behavior of the RCS of the target were recognized as additional sources of information to respond to issues of characterizing and classifying targets in the domain of radar [RIE 89].

As a scalar quantity, the RCS is a function of the polarization of incident and transmitted waves. A more complete description of the interaction between the incident wave and the tag is given by the polarimetric scattering matrix, also known as the Sinclair matrix [SIN 50].

An arbitrary-polarized plane wave can always be expressed as the superposition of two linear-polarized waves (horizontal and vertical for the sake of simplicity). The electrical field of a monochromatic wave that propagates along the axis $\vec{e}_z$, can be expressed like this:

$$\vec{E}^i = \vec{E}^i_h + \vec{E}^i_v = \left(E^i_h \vec{e}_h + E^i_v \vec{e}_v\right)e^{j(\omega t + kz)} \qquad [3.6]$$

The polarimetric scattering matrix relates the scattered electrical field $\vec{E}^s$ at the receiving antenna to the incident electric field $\vec{E}^i$, component by component. Using matrix notation, the polarimetric scattering matrix S is traditionally defined by:

$$\begin{bmatrix} E^s_h \\ E^s_v \end{bmatrix} = \begin{bmatrix} S_{hh} & S_{hv} \\ S_{vh} & S_{vv} \end{bmatrix} \cdot \begin{bmatrix} E^i_h \\ E^i_v \end{bmatrix} \qquad [3.7]$$

The terms S_{hh} and S_{vv} are *co-polarization* terms while the terms S_{hv} and S_{vh} are *cross-polarization* terms. Although it is often used in practice, definition [3.7] depends on the distance R between the tag and the antennas. One way to get away from this dependency is to factor the term related to the propagation of the spherical wave. This provides an alternative definition S' of the polarimetric scattering matrix:

$$\begin{bmatrix} E^s_h \\ E^s_v \end{bmatrix} = \frac{e^{-jkR}}{2\sqrt{\pi}R} \cdot \begin{bmatrix} S_{hh}' & S_{hv}' \\ S_{vh}' & S_{vv}' \end{bmatrix} \cdot \begin{bmatrix} E^i_h \\ E^i_v \end{bmatrix}, \qquad [3.8]$$

Because of their completely passive nature, chipless RFID tags are reciprocal. This means that for a monostatic scattering measurement, the scattering matrix is symmetrical, that is $S_{hv} = S_{vh}$. The elements of the matrix are complex and account for all phase changes caused by scattering. Once the scattering matrix of a chipless tag has been measured, it is possible to calculate the amplitude and polarization of the scattered wave, regardless of the polarization of the incident wave, simply by modifying the components of the vector $\vec{E}^i$ in [3.7]. In addition, it is possible to theoretically predict variations in the matrix S based on the orientation of the reader's transmitting and receiving antennas [RIE 89]. An example of an application will be provided in the next chapter.

Polarimetry has been researched extensively in the domain of radar. Some reference texts include the foundational works of Sinclair [SIN 50] or Kennaugh [KEN 52] as well as Huynen's thesis [HUY 70] and more recently the works of Boerner [BOE 81] or Cloude and Potier [CLO 96]. Recently, interesting applications of polarimetry have also appeared in the field of chipless RFID [RAN 16a, PÖP 16b].

3.1.3. *The electromagnetic signature of a chipless RFID tag*

According to its general definition, "the signature" is the set of elements that makes it possible to characterize the presence, type or identity of an object by a reading system.

A chipless RFID tag can be likened to a radar target but it still has particularities that are directly related to different types of applications. For applications where the cost of the tag is significant, chipless RFID tags are generally plane and their surface should not be larger than the size of a credit card because they must be able to be applied to packaging easily. This major constraint limits the RCS level to values in the order of −20 dBsm and read ranges are rarely greater than one meter [PER 14]. In the majority of cases, we are concerned with the response of the tag under normal incidence, which generally ensures an optimal energy coupling between the incident wave and the tag. In practice, however, it is common to use a bistatic measurement configuration to limit the coupling between antennas. In this case, relatively small angles are used (≤20°) in order to stay in the main lobe of the re-radiation pattern of the tag. Secondly, as noted, recent research has demonstrated that a higher coding density was obtained by frequency-coded tags. This implies that the signature of an RFID tag is broadband.

When only the magnitude of the response provides information, the "electromagnetic signature of a chipless RFID tag" will be used to designate the graphic representation of the RCS based on the frequency. When the polarization or the phase is significant, the "electromagnetic signature of the tag" will be used to designate the graphic representation (in amplitude and in phase) of one of the components of the scattering matrix based on the frequency. Depending on the case, we will focus more on the cross-polarization component S_{hv} or the co-polarization components S_{hh} and S_{vv}. If not otherwise indicated, the tag will be considered under normal incidence.

The electromagnetic signature is specific to a tag. It is easily accessible through simulation. For example, the simulation software *CST Microwave Studio* (CST) makes it possible to illuminate the tag using a plane wave and to position probes to measure the fields at a given distance. CST allows you to choose to either view the RCS directly (normalized in relation to the read range) or to view the field reflected at the probe, proportional to the matrix S as defined by [3.7]. On the other hand, the electromagnetic signature is more difficult to measure because it is an intrinsic quantity of the tag that must not take into account certain parameters related to measurement configuration, such as antenna gain or transmission channel properties. In the remainder of this chapter, we will see that it is possible to model a practical reading system in its entirety using the radar range equation. We will also see that a calibration phase is necessary to offset the errors that are likely to appear in measurements and to trace the electromagnetic signature.

3.2. Reading system

The reading system of a chipless RFID tag is very similar to impulse radar [GAR 16]. In laboratory conditions, we generally use a network analyzer (VNA) connected to the transmitting antenna in its port 1 and to the receiving antenna in its port 2. The measured quantity is the parameter S_{21} of the VNA which represents the ratio of waves measured (amplitude and phase) to the ports 1 and 2. It is still possible to consider low cost readers as in [GAR 16], but they generally have lower dynamic ranges.

3.2.1. *Radar range equation*

The radar range equation makes it possible to evaluate the overall performances of an RFID tag reading system.

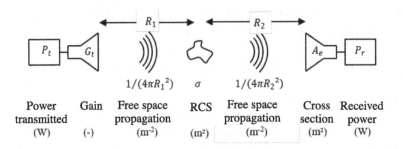

Figure 3.3. *Power budget of a radar link*

The simplest formulation of the radar range equation can be determined using the power budget between the transmitting and receiving antennas. Figure 3.3 illustrates the transfer of power from a transmitter to a tag and finally to the receiver. The radiating radar delivers a power P_t, radiated through an antenna with a gain G_t. At a distance R from the antenna, the power density W^i is simply the transmitted power modified by the antenna gain and divided by a factor $4\pi R_1^2$ associated with the attenuation of the spherical wave. Therefore, the power density can be expressed as:

$$W^i = P_t G_t / (4\pi R_1^2) \qquad\qquad [3.9]$$

The chipless RFID tag intercepts a part of the incident power. A fraction of this power is then radiated in the direction of the receiving antenna. As mentioned previously, the RCS is defined as the projection surface that is required to intercept and isotropically radiate an equivalent quantity of power. Theoretically, we can assume that the target captures the power P_i:

$$P_i = P_t G_t \sigma / (4\pi R_1^2), \qquad\qquad [3.10]$$

which is then isotropically radiated in space. The power density at the receiving antenna is therefore given by:

$$W^s = P_t G_t \sigma / (4\pi R_1^2 \cdot 4\pi R_2^2) \qquad\qquad [3.11]$$

The power received by the load of the receiving antenna is simply the product of the power density at the antenna and the effective aperture of the antenna, A_e. However, it is often more practical to work with the antenna gain that is expressed in relation to the effective aperture through the relation:

$$A_e = G_r \lambda^2 / 4\pi \qquad\qquad [3.12]$$

In addition, if we assume that the same antenna is used for transmitting and receiving, we have $G_t = G_r = G$ and $R_1 = R_2$. The power received is therefore given by:

$$P_r = \sigma \frac{P_t G^2 \lambda^2}{(4\pi)^3 R^4} \qquad\qquad [3.13]$$

This is the simplest form of the radar range equation that ignores a certain number of factors such as the polarization of the transmitting and receiving antennas. Nevertheless, this relation is extremely valuable for quickly assessing a reading system's performances. For example, the read range can be easily calculated by isolating the distance R in [3.13] and defining a minimum power threshold for the reception signal, P_{min}. The maximum detection range is therefore given by:

$$R_{max} = \left[\frac{P_t G^2 \lambda^2 \sigma}{(4\pi)^3 P_{min}} \right]^{1/4} \qquad\qquad [3.14]$$

The radar range equation models the overall detection system and makes it possible to predict the value of the module obtained directly by measuring the parameters S_{21} of a network analyzer. For example, for a monostatic measurement:

$$|S_{21}| = \sqrt{\frac{P_r}{P_t}} = \sqrt{\sigma} \frac{G\lambda}{\left(2\sqrt{\pi}\right)^3 R^2} \qquad\qquad [3.15]$$

This equation is determined using a power budget and it does not provide any information about the phase of parameter S_{21}. Information about the phase can be obtained but it requires knowing a detailed model of the behavior of the antennas.

3.2.2. Calibration

In practice, measuring the electromagnetic signature of a chipless tag generally requires a rigorous calibration phase. When the parameter S_{21} on the network analyzer corresponds to an empty detection environment measurement (without the tag, in an anechoic chamber), a level in the order of −40 to −50 dBm (for a transmitting power of 0 dBm) is observable. This response is linked to the direct coupling between the antennas and the possible reflections of static objects in the measurement environment. This value is often comparable or even greater than the power that is reflected by the tag (−60 to −70 dBm for a distance in the order of 0.5 m). Without calibration, the response is drowned out in the noise and no detection is possible. In addition, the identifiers of tags are coded at the level of the electromagnetic signature through slight variations in amplitude and phase. Therefore, the filtering effects of cables and antennas should not compromise the integrity of these distinctive elements.

The calibration method that was used in this study was described in detail and theoretically argued in [VEN 13c, WIE 91a, WIE 91b]. It is based on an empty measurement (without the tag), $S_{21}{}^{empty}$, and on the measurement of a reference element, $S_{21}{}^{ref}$, whose RCS, σ^{ref}, is known in advance. The tag whose signature we are seeking to determine is then measured ($S_{21}{}^{m}$). From these different measurements, it is possible to determine the RCS specific to the tag σ, using the equation:

$$\sigma = \left| \frac{S_{21}{}^{m} - S_{21}{}^{empty}}{S_{21}{}^{ref} - S_{21}{}^{empty}} \right|^{2} \cdot \sigma^{ref} \qquad\qquad [3.16]$$

The first step (numerator of [3.16]) consists of subtracting the result of the empty measurement $S_{21}{}^{empty}$, from that of the tag $S_{21}{}^{m}$. In a certain way, this comes down to isolating the response of the tag from its environment. The response obtained this way is then normalized with respect to the reference measurement (whole fraction). This makes it possible to eliminate the dependence on the read range and also to compensate for the filtering effect of the antennas. The whole fraction is then multiplied by the theoretical RCS of the reference σ^{ref} to recover the amplitude of the normalized response. This relation makes it possible to reach the measured RCS of the tag. A similar procedure can be implemented to measure components in the polarimetric scattering matrix. By taking the example of the cross-polarization measurement, the equation to use is:

$$S_{vh} = \frac{S_{21}{}^{m} - S_{21}{}^{empty}}{S_{21}{}^{ref} - S_{21}{}^{empty}} \cdot S_{vh}{}^{ref} \qquad\qquad [3.17]$$

S_{vh} and ${S_{vh}}^{ref}$ are the cross-polarized components in the polarimetric scattering matrix for the tag and the reference element respectively. They must not be confused with the parameter S_{21} of the network analyzer.

3.3. Re-radiation mechanisms for chipless tags

When a chipless tag is illuminated by an incident electromagnetic wave, the induced currents appear on the surface of the conducting structures. These induced currents are the source of the scattered field. The re-radiation pattern of the tag is determined by the amplitude and the phase of the induced currents. However, it is possible to distinguish two very different scattering modes for a chipless RFID tag.

3.3.1. *Structural mode and antenna mode*

The first component of the response of the tag is called the *structural mode* (Figure 3.4(a–c)). In a certain way, it corresponds to a specular reflection of a wave on the tag surface. If the incident wave is a pulse with a limited duration and a given polarization, this first part of the response is an image of the incident pulse (same duration and polarization). Because chipless tags have a planar structure, the component of the response related to the structural mode approximately follows the Snell–Descartes law of reflection, which means that we find the maximum power in the specular direction. The structural mode appears for a chipless tag but it is also the main scattering mode for most objects in the area. The "environmental response" also generally shares the same characteristics as the interrogation pulse: a broadband, limited-time (pulse duration) response with the same polarization.

The second scattering mode that appears for a chipless RFID tag is called the *antenna mode* (Figure 3.4(d–f)). It is related to the fact that the tags generally have a structure that is designed to capture part of the incident power and to guide it in particular direction through a transmission line. This mechanism is very easily illustrated by chipless tags that are time-coded (receiving antenna and transmitting antenna connected by a waveguide). The same type of behavior appears in frequency coded tags but it is used in a different way. For certain frequencies, there is a particular correspondence between the electrical length of the line and the wavelength, which creates a standing wave phenomenon on the line. These standing waves are typical for resonance. The power is no longer simply guided by the structure but stored in the tag's near-field. The energy is re-radiated in the whole space through a process that stretches out over time and depends on the quality factor of the structure. The distribution of induced currents along the transmission line determines the radiation characteristics of the tag such as the radiation pattern or the polarization of the transmitted wave.

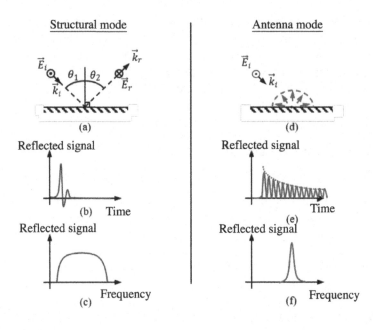

Figure 3.4. *Difference between the structural mode and the antenna mode for a chipless RFID tag coded in the frequency domain. The component of the response related to the structural mode (a-c) is an image of the incident pulse. It follows the Snell-Descartes law of reflection and preserves the same polarization as the incident pulse a). The response is limited in time b) and broadband c). For antenna mode (d-f), the polarization and the radiation pattern d) are determined by the distribution of currents along the line. The response is resonant, stretches out over time e) and presents a selective peak in the frequency f). For a color version of this figure, see www.iste.co.uk/rance/rfid.zip*

The total electromagnetic response of a chipless RFID tag is the sum of the fields related to the antenna mode and structural mode. It is important to note that in most cases, only the antenna mode is used to code information. This is true for both temporal tags (propagation delay along the line used to code information) and frequency tags (resonance).

3.3.2. Analogy with antennas

The structural and antenna modes were first identified during the characterization of antennas used as scatterers. As in the case of a chipless RFID tag, the structural mode appears because the antenna has a given shape, size and material. This mode is completely does not depend on the fact that the antenna was specifically designed to send the RF energy to a terminal load. On the other hand, the antenna mode corresponds to the fraction of the power that is captured by the antenna, guided toward the terminal load, reflected by impedance mismatch between the load and the line, and finally re-radiated by the antenna (Figure 3.5).

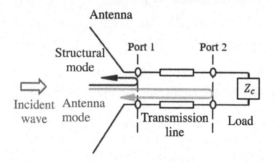

Figure 3.5. *Antenna used as a scatterer. The two modes*
of re-radiation are represented by a dark blue arrow (structural mode)
and a light blue arrow (antenna mode). For a color version of this
figure, see www.iste.co.uk/rance/rfid.zip

In the case of an antenna, it is fairly simple to isolate the contribution of each of the two modes by varying the terminal load of the antenna. There are different definitions of these modes depending on the reference load used for characterization as enumerated in [HAR 64, GRE 63, HAN 89]. The most common and most natural definition is that given by Green [GRE 63]. Green states that the structural mode is obtained when a transfer of maximum power occurs between the antenna and the load, which comes back to using a conjugate-matched load ($Z_L = Z_A{}^*$) at the terminals of the antenna. In this case the scattered field is noted as $E_s(Z_A{}^*)$. The scattered field for a given load $E_s(Z_L)$ can be expressed in relation to $E_s(Z_A{}^*)$ with the relation:

$$E_s(Z_L) = E_s(Z_A{}^*) + I(Z_A{}^*) \cdot \Gamma_A{}^* \cdot E_r \qquad [3.18]$$

where Γ_A^* is a modified reflection coefficient defined by:

$$\Gamma_A^* = \frac{Z_A^* - Z_L}{Z_A + Z_L}$$ [3.19]

In [3.18] we have:

$- E_s(Z_A^*)$: electrical field scattered by the antenna for a conjugate matched load;

$- I(Z_A^*)$: current induced by the incident wave at port 2 for a conjugate-matched load;

$- Z_A$: impedance of the antenna at port 2;

$- E_r$: electrical field radiated by the antenna when it is excited by a unit current. This term is expressed as an electrical field divided by a current.

The constant term $E_s(Z_A^*)$ of [3.18] was called the *structural mode* and it appears because although the antenna is conjugate-matched, there is always an induced current on the structure that creates a scattered field. The second term of [3.18] was called the *antenna mode* of the scattered field because it is determined solely from the radiation properties of the antenna and disappears when the antenna is conjugate-matched.

For a chipless RFID tag and in particular for the RF encoding particle (REP) approach that was introduced in the previous chapter, it is often more difficult to precisely characterize the antenna mode and the structural mode because these systems do not have ports properly speaking.

It has long been known that there is a quantitative relation between the gain and the RCS of an antenna due solely to its re-radiation properties. A simplified case study that does not consider the structural mode was proposed by Appel-Hansen [APP 79]. The idea is to consider an antenna in which all of the scattered energy comes from the energy collected by the aperture, which is propagated through a lossless transmission line and is re-radiated by the effect of a total reflection on the terminal load of the antenna. In relation to [3.18], it comes back to considering the particular condition: $E_s(Z_A^*) = 0$ and $|\Gamma_A^*| = 1$. In this case, the RCS considered is uniquely connected to the antenna mode (i.e. the term $I(Z_A^*) \cdot E_r$ in [3.18]) and will be noted σ_a.

Consider an antenna with an effective aperture A_e, illuminated by a plane wave with a power density W^i. The power that is collected by the antenna and transmitted toward the load is simply given by:

$$P_r = W^i \cdot A_e \qquad [3.20]$$

If the antenna is terminated by a purely reactive load, there is a total reflection and the re-radiated power is equal to the power received. For an antenna with a gain G, the power density re-radiated at a distance R is given by:

$$W^r = P_r G / (4\pi R^2) = W^i A_e G / (4\pi R^2) \qquad [3.21]$$

The definition of the RCS, expressed in terms of power densities is given by [3.3]. By replacing W^r in [3.3] with the expression [3.21], we get the expression of the RCS as a function of the antenna characteristics:

$$\sigma_a = A_e G \qquad [3.22]$$

which can only be expressed as a function of the gain by using the classic relation:

$$A_e = G\lambda^2 / 4\pi \qquad [3.23]$$

in such a way that:

$$\sqrt{\sigma_a} = \frac{G \cdot \lambda}{2\sqrt{\pi}} \qquad [3.24]$$

This result has the merit of being simple but it only accounts for one part of the energy scattered by the antenna. The scattering related to the structural mode was ignored.

3.3.3. *Application for the Design of REPs*

The equations [3.18] and [3.24] are of particular interest when the tag is designed according to the radiofrequency encoding particle (REP) approach [VEN 13b]. This approach consists of using a resonant element that carries out, by itself, the functions of receiving antenna, filter, and transmitting antenna. Based on the functions that these elements should provide, it seems clear that a resonant antenna used as a scatterer is a good candidate for a starting structure for this type of project.

We will examine some of the desired characteristics for an antenna used in the framework of REP design. First of all, we saw that the mode used for coding was the antenna mode. The structural mode that appeared in [3.18] can therefore be seen as additive noise that depends on the measurement configuration and is likely to disrupt detection. Consequently, it is preferable that the antennas used have a low structural mode, as in the case of microstrip antennas. According to [3.24], the scattered field associated with the antenna mode is directly proportional to the antenna gain. The result makes it possible to estimate the RCS level of the tag based on the antenna used. In addition, knowing the gain based on the incidence angle also makes it possible to predict the re-radiation pattern of the tag. From an application standpoint, a directional antenna is not necessarily preferable because the RCS level of the antenna depends strongly on the incidence angle and the observation angle.

Equation [3.18] also provides information about the type of load to consider to maximize the component related to the antenna mode. If we consider a purely reactive load, it is clear that $|\Gamma_A^*| = 1$, so that all of the guided power is reflected at the load. The power may be stored in the load but no power dissipates. This result is also reached for the particular case of a short-circuit ($Z_L = 0$) or open-circuit ($Z_L = +\infty$) antenna.

With the RF encoding particle approach, the concept of load is less clear than with a traditional antenna because it is not possible to define a port on its own. In most cases, however, the encoding particles are inspired by traditional planar antenna structures (Figure 3.6(a–c)) for which it is possible to achieve particular load conditions without recourse to a lumped element. For example, a dipole (Figure 3.6(a)) can be used as a REP if the terminals of the dipole are connected by a short-circuit. Contrary to what we might expect from [3.24], the radiation pattern of this type of scatterer is different from that of a dipole used for transmitting due to the large structural mode [HAN 90]. Similarly, a CPS line has high radiation losses when one of the extremities is in open circuit and can therefore be treated as an antenna. This kind of line can be used as a REP (Figure 3.6(b)) if the other extremity is short-circuited. This case will be studied in detail in the next chapter. A rectangular microstrip patch antenna (Figure 3.6(c)) can also be used as a REP by considering an open circuit at the port. It is possible to use a zero-length feeding line to obtain a more compact device. In this case, the REP re-radiation pattern is very similar to the patch radiation pattern due to the very weak structural modes of the microstrip structures. The equation [3.24] is therefore an accurate model of the behavior of a patch used as an REP.

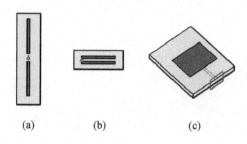

(a)   (b)   (c)

Figure 3.6. *Example of antenna structures used for REP designs; a) Dipole; for the REP, the dipole terminals are connected by a short-circuit; b) CPS line; for the REP (C-resonator), one of the extremities of the line is short-circuited; c) Rectangular patch antenna; for the REP we consider an open-circuit termination. It is also possible to use a zero-length feeding line. For a color version of this figure, see www.iste.co.uk/rance/rfid.zip*

The parallel between the antenna structures and the associated resonant scatterers makes it possible to use the traditional results of antennas for the first step of the design. For example, for an REP based on a microstrip dipole, it is possible to determine the resonant frequency by referring to the traditional relations in the literature [JAC 09]. Similarly, [3.24] explains certain scatterer behaviors thanks to the characteristics of associated antennas. Returning to the example of the microstrip dipole, it is possible to predict the evolution of the RCS as a function of the dielectric losses in the substrate analytically. The gain can be expressed as a function of the radiation efficiency e_r and the directivity D of the patch antenna:

$$G = e_r \cdot D \tag{3.25}$$

With resonance, the antenna behaves like a parallel RLC circuit [BAL 05] and the radiation efficiency can be expressed as a function of the quality factor associated with the radiation, Q_r, compared to the total quality factor, Q_{tot}, by:

$$e_r = \frac{Q_{tot}}{Q_r} \tag{3.26}$$

For a patch, the field is concentrated in the cavity formed by the ground plane and the upper metallic face of the antenna, and the losses are dominated by dielectric losses. As a first approximation, the total quality factor can be expressed as:

$$\frac{1}{Q_{tot}} = \frac{1}{Q_r} + \frac{1}{Q_d}$$
[3.27]

where Q_d is the quality factor related to dielectric losses. It is known for patches [JAC 09] that:

$$Q_d = \frac{1}{tan\,\delta}$$
[3.28]

Equations [3.24]–[3.28] make it possible to express the RCS of the encoding particle as a function of substrate losses:

$$\sqrt{\sigma} = \frac{1}{1+Q_r\cdot tan\delta} \cdot D\lambda/\left(2\sqrt{\pi}\right)$$
[3.29]

We have completed a series of simulations of a resonant microstrip dipole at 3.92 GHz by varying the loss tangent of the substrate. The patch is oriented at 45° (Figure 3.7) and the measurement is realized in cross-polarization in order to remove the component of the response related to the structural mode (the polarizations of the transmitting and receiving antennas are indicated by the arrows in Figure 3.7). The simulated response corresponds to the field reflected by a probe located at a distance of 1 m from the target. The amplitude of the peaks of the response is compared with the amplitude predicted by the model established using the properties of the patch antenna [3.29]. The term $D\lambda/\left(2\sqrt{\pi}\right)$ is obtained by simulation fitting for the case where there are no dielectric losses (tan δ = 0). We observe a perfect correlation between the simulation results and the theoretical prediction.

This relatively simple study clearly illustrates the value of establishing an antenna-based model for designing an REP. The traditional properties of antennas can then be used to establish the link with the amplitude of the RCS obtained. It should be noted that, unlike traditional antennas, REPs are not intended to guide energy toward a particular load because ideally, all of the power is re-radiated. The location of "fictitious" ports for an REP is arbitrary and can be positioned wherever is most suitable for analyzing the structure.

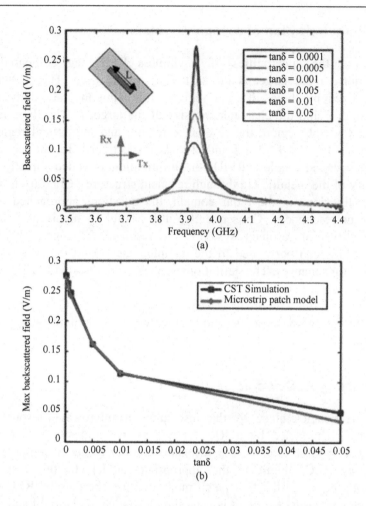

Figure 3.7. *Results of the simulation (CST) of the REP microstrip patch in the form of a dipole that is L = 20 mm long and W = 2 mm wide. The substrate used in the simulation has a thickness of t = 0.8 mm and a permittivity of ε_r = 3.38 that corresponds to that of the RO4003. We simulated the variation of dielectric losses in the substrate and the simulated result is compared with the one predicted in theory; a) Response of the simulated tag near resonance (cross-polarization); b) Comparison of the amplitude of the simulated peak with the amplitude calculated according to the equation [3.29]. For a color version of this figure, see www. iste.co. uk/rance/rfid.zip*

3.4. Characterization of resonant systems

One of the specificities of a chipless RFID tag compared to a conventional radar target is the scattering regime used. The scattering is characterized by three different regimes following the ratio between the wavelength λ and the characteristic size of the target L. The three regimes are the Rayleigh region, the resonance region, and the optical region that correspond to $\lambda \gg L$, $\lambda \propto L$ and $\lambda \ll L$ respectively. In the context of frequency coded chipless RFID, the information is coded directly in the geometry of the metallic tag through resonant structures. Contrary to what is traditionally done in the radar domain, the questions approached refer to significant resonances. Consequently, it is useful to precisely define the characteristics of resonant systems: the resonant frequency, the quality factor, which can be related to the damping factor and the bandwidth for significant resonances. The definitions of these quantities will be given from an energy standpoint (general) but will be illustrated for the case of an RLC series circuit which will serve as a simple study. RLC circuits are used in a very traditional way to model the response of a resonant chipless RFID tag.

3.4.1. Series RLC circuit

At frequencies close to the resonance, a microwave resonator can generally be modeled by a RLC circuit, so we will study some of the properties of these circuits. For resonance similar to the first dipole mode, the series RLC circuit is the appropriate model. In the case of an antiresonance, as in the first patch mode, we use the parallel RLC circuit. Studying a series RLC circuit is very similar to a parallel circuit and we will only discuss the series configuration. Study of a parallel circuit applied to electromagnetic resonators is done in [COL 01, POZ 12]. These two reference works also address the case of distributed electromagnetic resonators.

Here, we will study the harmonic response of the series RLC resonant circuit that is represented in Figure 3.8.

The input impedance of a circuit like this is:

$$Z_{in} = R + j\omega L - j\frac{1}{\omega C} \qquad [3.30]$$

and the complex power delivered to the resonator is:

$$P_{in} = \frac{1}{2}VI^* = \frac{1}{2}Z_{in}|I|^2 = \frac{1}{2}Z_{in}\left|\frac{V}{Z_{in}}\right|^2 = \frac{1}{2}|I|^2\left(R + j\omega L - j\frac{1}{\omega C}\right) \quad [3.31]$$

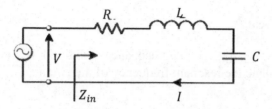

Figure 3.8. *Series RLC circuit*

The power dissipated by the resistance is:

$$P_{loss} = \frac{1}{2}R \cdot |I|^2 \qquad\qquad [3.32]$$

The average magnetic energy stored by the inductor L is:

$$W_m = \frac{1}{4}L \cdot |I|^2 \qquad\qquad [3.33]$$

The average electric energy stored by the capacitor C is:

$$W_e = \frac{1}{4}C \cdot |V_c|^2 = \frac{1}{4}\frac{1}{\omega^2 C}|I|^2 \qquad\qquad [3.34]$$

where V_c is the tension applied to the terminals of the capacitor. This means that the complex power of [3.31] can be written in the form:

$$P_{in} = P_{loss} + 2j\omega(W_m - W_e) \qquad\qquad [3.35]$$

And the input impedance [3.30] can be rewritten in the form:

$$Z_{in} = \frac{2P_{in}}{|I|^2} = \frac{P_{loss} + 2j\omega(W_m - W_e)}{\frac{1}{2}|I|^2} \qquad\qquad [3.36]$$

Although established using the RLC series circuit, this equation is valid for all resonant circuits with a single port if the terminal current I can be correctly defined. Based on this definition, the resonance condition

$W_m = W_e$ corresponds to a purely resistive input impedance. In the case of an RLC circuit, based on [3.36] and [3.32], the input impedance is given by:

$$Z_{in} = \frac{P_{loss}}{\frac{1}{2}|I|^2} = R \qquad [3.37]$$

which is purely real.

Based on [3.33], [3.34], the resonance condition $W_m = W_e$ makes it possible to calculate the resonant frequency of the circuit:

$$\omega_0 = \frac{1}{\sqrt{LC}} \qquad [3.38]$$

3.4.2. Quality factor

An important parameter that characterizes the frequency selectivity and the general performances of a resonant circuit is the quality factor, Q. A general definition of Q that applies to all resonant systems is:

$$Q = \omega_0 \frac{W}{P_{loss}} \qquad [3.39]$$

where W represents the average energy stored by the system and P_{loss}, represents the energy lost per second. The average energy stored by the system is divided between the electrical and magnetic fields, which gives:

$$Q = \omega_0 \frac{W_m + W_e}{P_{loss}} \qquad [3.40]$$

This means that Q can be seen as a measurement of the losses of a resonant circuit.

In the case of an RLC series circuit, the resonator losses are represented by the resistance R. The quality coefficient Q_0 of the total structure is obtained by considering the resonant series circuit in Figure 3.8. Q_0 can be expressed using the parameters of the circuit from the equations [3.40] and [3.32]–[3.34] and the resonance condition $W_m = W_e$:

$$Q_0 = \omega_0 \frac{2W_m}{P_{loss}} = \frac{\omega_0 L}{R} = \frac{1}{\omega_0 RC} = \frac{1}{R}\sqrt{\frac{L}{C}} \qquad [3.41]$$

which shows that generally Q increases when losses decrease.

Contrary to resonators used to produce filters, radiation losses are desirable in the case of chipless RFID because they translate the tag's capacity to radiate. It is often useful to separate the different sources of losses in the equivalent RLC circuit by different resistances positioned in a series. The types of losses that are generally encountered are radiation losses R_r (desired), dielectric losses R_d and conduction losses R_c. It is possible to associate a quality factor to each of these losses:

$$Q_r = \frac{\omega_0 L}{R_r} \qquad\qquad Q_d = \frac{\omega_0 L}{R_d} \qquad\qquad Q_c = \frac{\omega_0 L}{R_c} \qquad\qquad [3.42]$$

Based on the definitions [3.41] and [3.42], we can see that the total quality factor of the structure is given by:

$$\frac{1}{Q_0} = \frac{1}{Q_r} + \frac{1}{Q_d} + \frac{1}{Q_c} \qquad\qquad [3.43]$$

Generally, for a chipless tag, the presence of dielectric losses and conduction losses decrease the overall quality factor of the structure without improving the tag's capacity to radiate. This is a major problem for tags printed on low cost substrates like paper because the dielectric losses are very significant. In this case, the overall quality factor is not dominated by Q_r (associated with radiation) but by the dielectric losses of the substrate.

3.4.3. Damping factor

Another important parameter related to resonant circuits is the damping factor, ξ. This parameter measures the rate of decrease of oscillations when the source is removed. For a circuit with an elevated quality factor, ξ can be evaluated with Q using the perturbation method. When the losses are present, the energy stored in the resonant circuit decreases proportional to the average energy present at any time (because $P_{Loss} \propto II^*$ and $W \propto II^*$, we have $P_L \propto W$), so that:

$$\frac{dW}{dt} = -2\xi W \qquad\qquad [3.44]$$

which can be written in the form:

$$W = W_0 e^{-2\xi t} \qquad\qquad [3.45]$$

where W_0 is the energy present in the system at $t = 0$. The rate of decrease of W must be equal to the power lost, so that:

$$-\frac{dW}{dt} = 2\xi W = P_L \tag{3.46}$$

As a result, using the definition of Q [3.39] we have:

$$\xi = \frac{P_L}{2W} = \frac{\omega_0}{2}\frac{P_L}{\omega_0 W} = \frac{\omega_0}{2Q_0} \tag{3.47}$$

The damping factor is inversely proportional to the quality factor Q. Instead of [3.45], we can write:

$$W = W_0 e^{-\omega_0 t/Q_0} \tag{3.48}$$

This relation is interesting in the context of chipless RFID because it makes it possible to evaluate the quality factor of a resonator based on its time response (on the condition of having a single resonance). However, it is easier to use the linearized form of [3.48]:

$$\ln(W) = \ln(W_0) - \omega_0 t/Q_0 \tag{3.49}$$

Generally, chipless tags have several resonances and it is therefore not possible to use [3.49] directly. The method used during this study to evaluate the quality factor(s) of resonators is based on the parametric identification of the frequency response with that of a second order band-pass filter (RLC circuit). The frequency representation makes it possible to separate the resonances (disjointed peaks) more easily than a time representation.

3.4.4. Bandwidth

Another parameter that is characteristic of the resonance is the 3 dB bandwidth (BW). We can consider the behavior of the input impedance of the resonant circuit in the vicinity of the resonant frequency: $\omega = \omega_0 + \Delta\omega$. The input impedance can be rewritten in the form:

$$Z_{in}(\omega) = R + j\omega L \left(1 - \frac{1}{\omega^2 LC}\right) = R + j\omega L \left(\frac{\omega^2 - \omega_0^2}{\omega^2}\right) \tag{3.50}$$

Now, $\omega^2 - \omega_0^2 = (\omega - \omega_0)(\omega + \omega_0) = \Delta\omega(2\omega - \Delta\omega) \simeq 2\omega\Delta\omega$ for the low values of $\Delta\omega$. We can therefore carry out a first-order approximation of the impedance in the vicinity of the resonance:

$$Z_{in}(\omega_0 + \Delta\omega) \simeq R + j2L\Delta\omega = R + j\frac{2RQ_0\Delta\omega}{\omega_0} \qquad [3.51]$$

Figure 3.9 represents the variation of the input impedance module based on the frequency for an RLC series circuit.

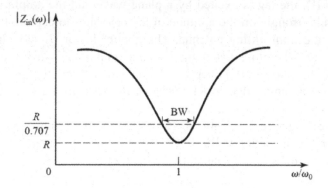

Figure 3.9. *Input impedance module with respect to frequency*

Using the voltage expression in [3.31] we can see that the *real* average power delivered to the circuit is half of the power delivered at the resonance for the impedance condition $|Z_{in}|^2 = 2R^2$. If the BW is defined as the fractional bandwidth, the upper limit of the band is obtained by $\Delta\omega/\omega_0 = BW/2$. By using this relation in [3.51] as well as the impedance condition, we find:

$$|R + jRQ_0BW|^2 = 2R^2 \qquad [3.52]$$

which can be simplified as:

$$BW = \frac{1}{Q_0} \qquad [3.53]$$

This relation is relevant for chipless tags because it makes it possible to evaluate the frequency band occupied by a peak and then define the size of the frequency windows to use for frequency coding.

3.4.5. *Electromagnetic resonators*

We have defined the main characteristics of resonators. Resonant circuits are very commonly used for RF components such as oscillators, amplifiers, frequency filters, and so on. The main difference between these circuit applications and the use of resonant scatters in the context of RFID is primarily related to the fact that in the first case, the resonators are an integral part of a more complex RF system and are excited by dedicated techniques such as inductive coupling with a feed line. In the context of chipless RFID, the tag is excited by a plane wave and the amplitude of the RCS depends strongly on the aptitude of the resonator to capture the energy emitted by the transmitting antenna. The quality factor Q_r associated with radiation losses therefore has a fundamental importance in the context of chipless RFID while it is generally considered a term of pure losses for resonant circuits embedded in RF systems. It is interesting to compare the different technologies used to create resonators in other domains based on the frequency range of applications. Table 3.2 provides a quick classification of the technologies addressed in [COL 01] with their frequency ranges and the indicative values of associated quality factors.

Frequency range	> 100 MHz	100 MHz–1000 MHz	> 1000 MHz	
Technologies	Circuits with lumped elements (RLC)	Resonant circuits based on transmission lines	Microwave cavities	Dielectric resonators
Typical Q	–	From several hundred to around 10 000. Microstrip resonators: between 200 and 600.	Rectangular cavity: @7 000 MHz; Q ≃ 12 000.	From 100 to several hundred.
Comments	Too many losses for higher freq.	At frequencies greater than 1000 MHz, transmission line resonators have Q values that are relatively low and it becomes preferable to use cavities.	–	Reduced sizes (around one tenth) compared to cavities at the same frequency.

Table 3.2. *Comparison between several technologies for resonant circuits and their quality coefficients*

On the frequency range used in chipless RFID (UWB, from 3.1 GHz to 10.6 GHz), it appears fairly clear that the resonant cavities have the highest Q values and that they are more desirable from a performance standpoint. However, it seems difficult to conceive using a cavity while keeping the tag low cost and potentially printable. Although they generally have lower quality factors, technologies such as resonant circuits based on transmission lines and microstrip resonators are planar and therefore easily adaptable as REPs.

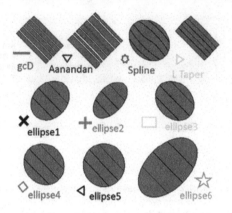

Figure 3.10. *Shapes of microstrip resonators simulated to study the influence of the shape on the quality factor. For a color version of this figure, see www.iste.co.uk/rance/rfid.zip*

A study of microstrip resonators shapes (coupled dipoles) was carried out. The objective was to see to what extent it was possible to adjust the quality factor of a given technology by varying the resonator's shape. A basic scatterer with five uniformly-spaced microstrip dipoles with identical dimensions was used as a reference case (*gcD* for gap-coupled Dipoles in Figure 3.10). The width of these dipoles is 2 mm and the spacing between the dipoles is 0.5 mm. The dipoles are positioned on an RO4003 type substrate ($\varepsilon_r = 3.38$) with a thickness of 0.8 mm with a ground plane. Simulations are carried out for different lengths of gap-coupled dipoles in order to consider the impact of the resonant frequency on the quality factor, the red curved line in Figure 3.11. The objective is to see if by varying the size or the shape of different dipoles within the same REP, it is possible to modify the quality factor and to obtain a larger band response than in the uniform case. Resonators with variable forms are simulated (see Figure 3.10) and the corresponding quality factors are extracted by fitting with a second order resonator (series RLC circuit). In all cases, the dipoles are oriented at 45° and the measurements are made in cross-polarization in order to limit the contribution of the structural mode. In order to consider the influence of the resonant frequency, the quality factor is represented as a function of the frequency in Figure 3.11. We can see that for microstrip technology, the value of the quality factor depends on the resonant frequency of the REP, but very little on the form factor.

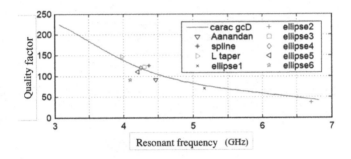

Figure 3.11. *Quality factor as a function of resonant frequency for microstrip resonators with variable forms. For a color version of this figure, see www.iste.co.uk/rance/rfid.zip*

3.5. Separation of the tag and its environment

We noted at the start of this chapter that the response of a chipless tag is generally lower than that of the environment and that a calibration procedure (empty measurement and reference measurement) is necessary to measure the tag's signature. However, there are techniques that make it possible to improve the readability of the tag compared to its environment. These methods rely on using specific characteristics of the tag's antenna mode which is used in the vast majority of cases to code the identifier. It is noteworthy that these techniques also generally make it possible to separate the contribution of the tag's response related to antenna mode from that of the structural mode.

3.5.1. *Depolarizing tag interrogated with cross-polarization*

The first approach is based the ability to control the polarization of the tag's response. For the antenna mode, a part of the power captured by the tag is guided within the structure. By varying the form and orientation of the metallic structure, it is possible to create induced currents with a particular direction. For example, the distribution of induced surface currents on a dipole oriented at 45° and illuminated by a vertically polarized plane wave is represented in Figure 3.12. It is known in the domain of antennas that in the far-field, the radiation pattern and polarization of an antenna can be calculated like the Fourier transform of the distribution of currents on the structure. In the case of the dipole represented in Figure 3.12, the polarization of the reflected wave for the antenna mode (indicated by a black arrow in Figure 3.12) follows the principal direction of the dipole. Although the dipole is excited by a vertically polarized plane wave, the reflected wave has two components, one polarized vertically (blue arrow in Figure 3.12) and the other polarized horizontally (red arrow in Figure 3.12). Using a horizontally polarized receiving antenna, it is possible to only select the horizontal component of the dipole's response.

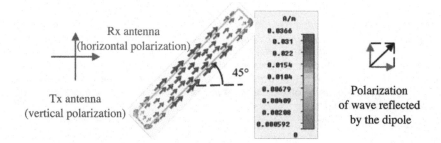

Figure 3.12. *Distribution of surface currents for a dipole illuminated by a vertically polarized incident wave. The field reflected by the dipole is polarized according to the main direction of the dipole (45°). It therefore has a vertically polarized component and a horizontally polarized component. The latter is detected by the receiving antenna. For a color version of this figure, see www.iste.co.uk/rance/rfid.zip*

This type of behavior is common for resonant objects but it does not appear for classical objects for which the reflected wave has typically the same polarization as the incident wave (see Figure 3.13). By associating a resonant tag oriented suitably with a cross-polarization reading, it is possible to isolate the tag's response from that of its environment. An example of an application of this approach is the implementation of chipless tags that can be decoded when placed on metallic objects such as canned goods [VEN 13c].

As we saw previously, the total response of the tag is the complex summation of the component of the field related to the structural mode (broadband, not used for coding) and that of the antenna mode (resonant, polarized in a specific direction). Using the same process as before, it is possible to separate these two modes in order to isolate only the structural mode used to code the information.

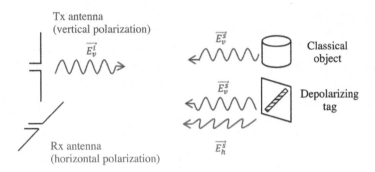

Figure 3.13. *Depolarizing tag positioned in the vicinity of a classical object. If the tag is measured in co-polarization, the response of the object will be added to that of the tag, which can lead to reading errors. In cross-polarization, only the horizontal component of the tag is detected, which makes it possible to isolate the response of the tag from the response of objects around it. For a color version of this figure, see www.iste.co.uk/rance/rfid.zip*

As an example, Figure 3.14 represents the co-polarization and cross-polarization responses of a rectangular microstrip patch oriented at 45°. The lateral dimensions of the patch are $L = 20$ mm and $P = 12$ mm. The patch is realized on a RO4003 ($\varepsilon_r = 3.38$) type substrate that is $t = 0.8$ mm thick and has a ground plane. In co-polarization, the response of the patch is the sum of the structural mode (mainly the broadband response of the ground plane) and the resonant response of the antenna mode. At the resonant frequencies of the patch (determined by the size of the lateral dimensions L and P), these two components are out of phase and dips appear at the level of the overall response. The depth of the dips is relatively small in relation to the average level of the response. In cross-polarization, only the resonant response (antenna mode) is visible on the signature. The height of the peaks is comparable to the depth of the dips but the main difference is that the height of the peaks is much greater compared to the non-resonant parts of the response. There is a shift of about 10 MHz between the frequencies of the associated peaks and dips. This difference can be simply explained by the fact that the presence of a dip is related to the summation between the resonant signal and another component. The maximum of the sum appears at a different frequency from the maximum of the resonant component.

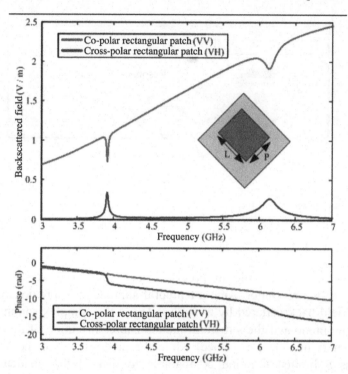

Figure 3.14. *Electromagnetic signature of a rectangular microstrip patch in co-polarization and cross-polarization. For a color version of this figure, see www.iste.co.uk/rance/rfid.zip*

From this example, we can see that it is much easier to identify the characteristics of resonances from the response in cross-polarization because it consists of traditional resonance figures.

3.5.2. *Temporal separation*

When we directly examine the spectral response of a chipless RFID tag, it is sometimes difficult to identify the resonance if it is low compared to the response of the substrate. The example of the signature of a CPS resonator without a ground plane in the form of a loop (CPS line short-circuited at each of its extremities) [RAN 16b] is represented in Figure 3.15 for a co-polarization measurement. This tag resonates between 6.5 and 7 GHz but it is difficult to determine the resonant frequency precisely.

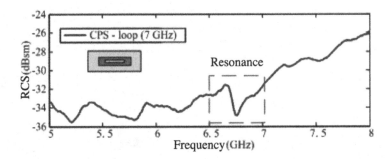

Figure 3.15. *Response of a CPS resonator in the form of a loop. The amplitude of the resonance is low compared to the component of the structural mode. It is difficult to gather information related to the resonant frequency. For a color version of this figure, see www.iste.co.uk/rance/rfid.zip*

However, it is possible to precisely determine the natural resonant frequency of the resonator using a temporal approach. As previously noted, the total field backscattered by a chipless tag is the complex summation of the antenna mode and the structural mode. The contribution of the antenna mode extends in time (resonance) while the contribution of the structural mode has a limited duration comparable to that of the incident pulse. Consequently, we can distinguish two different regions at the level of the temporal response of the tag. The "early time" of the response is composed of the summation of the contribution of the structural mode and the antenna mode. Given the specular nature of the reflection of the substrate, the "early-time" is very dependent on the variation of incidence or observation angles. The region of "late-time" occurs after the extinction of the specular reflection. The response is composed entirely of the natural resonances of the tag. The theory associated with the Singularity Expansion Method (SEM) [BAU 80, REZ 15a] demonstrated that in this region, the response of any scatterer (in terms of fields) can be described using the sum of damped sinusoids:

$$r(t) = \sum_{i=1}^{N} a_n e^{-\xi_n \cdot t} \cos(\omega_n t + \phi_n), \quad t > T_L \qquad [3.54]$$

where a_n and ϕ_n are the amplitude and the phase of the nth mode that depends on incidence and observation angles (aspect angles) and the form of excitation. ξ_n and ω_n are the damping factor and the pulsation of the nth mode, respectively. The quantity $\xi_n + j\,\omega_n$ is generally called the complex natural resonance of the nth mode. T_L is the start of "late-time". A major result from SEM theory is that it demonstrated that the natural resonances of any radar target do not depend on the form of the incident wave nor the aspect angles (incidence/observation). The set of natural frequencies is unique for a specific target and provides an interesting base to identify the target in the domain of radar [BAU 91, BAU 06]. In the context of chipless RFID, simple scatterers such as a loop resonator or a C-resonator can be assimilated to a single scattering center, and the summation [3.54] reduced to a single term. In this case, the natural resonant frequency described by SEM theory corresponds to the resonant frequency associated with the antenna mode.

In order to illustrate the relevance of this approach for chipless RFID, we can view the spectrogram of a loop resonator (Figure 3.16) calculated on the basis of the response measured. The spectrogram is obtained either from the temporal response or the frequency response by applying the Short Time Fourier Transform (STFT) [REZ 15a]. Here the local function (in time and frequency) used as window function is a chirp, because its form is quite similar to that of an incident pulse after being deformed by the antennas.

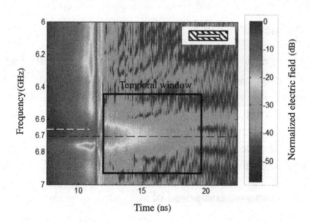

Figure 3.16. *Spectrogram obtained from the measurement of a CPS tag in loop form. The "early-time" corresponds to the zone below 12 ns. The "late-time" ranges from 12 to 20 ns and only contains the resonant components of the response. For a color version of this figure, see www.iste.co.uk/rance/rfid.zip*

The specular reflection due to the dielectric substrate is observable in the region below 12 ns. The resonance lasts longer than the specular reflection (up to 20 ns). During the first part of the response ("*early-time*"), the response of the resonator is comparable to that of the dielectric substrate and a shift in the resonant frequency (referred to by white dots on the spectrogram) is observed because of this additional contribution. A method to determine the natural resonance is to isolate the resonant part ("*late-time*") from the specular part ("*early-time*") by applying a temporal (and frequency) windowing of the signal [RAM 16b]. The temporal window is represented by a black rectangle on the spectrogram. It is chosen in order to preserve only the resonant part of the signal. In this way, the natural resonant frequency can be recuperated. It is possible to get back to the frequency or time domain by applying the inverse STFT but the amplitude of the peaks obtained does not correspond to the initial amplitude given that a part of the power contained in the response was eliminated by the windowing. An example of an application of this extraction method for chipless RFID tags including a discussion of the type and size of windows used is given in [RAM 16b].

3.6. Conclusion

This chapter explained the main characteristics of the response of chipless RFID tags. It is useful to briefly review the main results presented in this chapter:

– What is the measurable quantity?

In a case where only amplitude is important, we generally use the RCS to characterize the response of a chipless tag. If we want to find information about the phase or the polarimetric behavior of the tag, we use the polarimetric scattering matrix. For frequency tags, the electromagnetic signature is the representation of one of these quantities with respect to the frequency. In most cases, a calibration step is necessary to determine these quantities at the time of measurement.

– What is this response composed of?

The response of a chipless tag is composed of two different kinds of contributions. First, there is the structural mode, generally broadband and limited in time, which is not used for coding. The second component is the antenna mode which is selective in frequency and extends in time. This mode is used to code information on a frequency tag.

– How can the response be characterized?

The use of resonators causes peaks to appear in the signature of the tag. These peaks are characterized by their frequency and quality factor. For strong resonances, it is possible to relate the quality factor to other quantities such as the damping factor or the bandwidth.

– How can the response be isolated in a noisy environment?

The resonant nature of the antenna mode makes it possible to isolate the useful part of the tag's response from its environment. It is possible to use a depolarizing tag associated with a cross-polarized reading and also to temporally separate the components of the response.

The different theoretical notions addressed in this chapter will be used in the following chapters for the design of chipless tags. Chapter 4 in particular will show that by varying the polarimetric behavior of the resonators, it is possible to control the amplitude of the tag's response.

Magnitude Coding

4.1. Introduction

Coding information is a major issue in chipless RFID, where coding capacity is still low compared to current identification technologies like barcodes and traditional RFID. Increasing the coding capacities of tags is the condition that will allow chipless technology to become a real alternative. A lot of work has been done in the past few years to significantly increase coding capacity and some tags can now exceed fifty bits [VEN 12a], which is equal to the coding capacity of EAN13 barcodes. However, the market for tags that can only contain a small amount of information remains extremely limited. For example, we estimate that to break into the mass market the memory capacity of tags must be at least 128 bits [HAR 10], which corresponds to the quantity of information necessary to implement an Electronic Product Code (EPC), the global standard of reference. For industrial use, the size of the tag must not exceed the size of a credit card (8.5 cm × 5.4 cm).The objective of 128 bits corresponds to a surface coding density of 2.8 bits/cm².

Magnitude coding is a hybrid coding method that is directly aligned with the issue of increasing the quantity of information coded by the tag. The general idea is to code information in the frequency as well as in the magnitude of the response. By increasing the number of bits coded per resonator, this approach could make it possible to reach higher coding densities. Beyond coding considerations, the magnitude of the RCS is an important factor in itself in the context of chipless RFID because it determines the read range of the overall system.

Magnitude coding is also an important intermediary step to approach RCS synthesis. The method chosen for the synthesis amounts to project the objective onto the base of resonators' response. It is therefore necessary to carry out a certain number of operations on the elements of the base. The most important operation is without a doubt the multiplication by a scalar, which physically comes down to changing the magnitude of the elements. It is therefore necessary to develop practical techniques to control the magnitude of each element separately. This is the central issue of magnitude coding and the techniques developed in this chapter will be revisited in the next chapter which focuses on RCS synthesis.

This chapter also addresses the difficulties encountered by tags where the magnitude related to each resonator is controlled at both the level of the design and the level of measurement. The RCS magnitude is a sensitive parameter that depends on the nature of the excitation (type of pulse, orientation, polarization of incident wave) as well as parameters directly related to the measurement configuration, for example the tag-reader distance or the properties of the transmission channel.

4.1.1. *Hybrid coding*

In the beginning, most coding techniques for chipless RFID used a single physical quantity and only two possible states that typically corresponded to the absence or presence of a resonator [PRE 09c, REZ 12]. The most traditional example is frequency coding, where each bit coded corresponds to the presence or absence of a peak at a given frequency on the spectrum. This type of coding is represented schematically in Figure 4.1(a). Because each peak is associated with a physical resonator, there is a coding efficiency of 1 bit per resonator. Although not very powerful, this coding has the advantage of being simple and robust due to the strong contrast between the two possible states of each symbol. For size reasons however, we can only incorporate a limited number of resonators in a tag the size of a credit card and it is therefore not possible to reach 128 bits with this type of coding. For this approach, a maximum capacity of about 20 bits can be reached with an optimized tag [VEN 11a].

A significant increase in coding capacity can be reached if each resonator codes several different states. Various examples of the use of this idea were given in Chapter 2. The simplest case is when the physical quantity used for coding can take several different values. The variable quantity must be associated with a geometric parameter of the resonator.

It is generally accepted in the field of chipless RFID that the best coding density is obtained with *Frequency Position Coding* (FP), the same type as presented in Figure 3.1(b). In the case of frequency position coding, the frequency at which the peak appears is related to the length of the resonator. For a variable length, we can consider that the peak can take discreet values corresponding to $F_0 + k \cdot \delta F$, where δF is the frequency resolution. The frequency resolution δF characterizes the smallest separation in frequency that it is possible to distinguish when measuring between two successive peaks. In practice, δF depends on the selectivity of the resonator, the accuracy of the design and the application conditions. Perturbations related to the environment at the time of reading can have a significant influence on the response and determining the resolution in realistic conditions is necessary.

If we have a frequency range ΔF for coding, the number of different states that the resonator can code is $N = \Delta F / \delta F$. In this case, the coding efficiency is therefore $\log_2(N)$ bits per resonator.

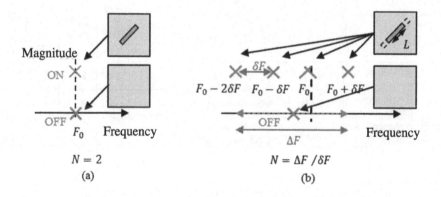

Figure 4.1. *Frequency coding: a) coding in presence/absence; b) frequency position and presence coding. For a color version of this figure, see www.iste.co.uk/rance/rfid.zip*

To increase the number of states coded by each resonator even more, an approach called hybrid coding can be used. Hybrid coding consists of combining several different physical quantities to code information. Each quantity must be controlled by independent geometric parameters to get the maximum benefit of this approach. Since frequency position coding is easy to implement and performs well in terms of coding capacity, it is often associated with a second physical quantity. There are several examples in the literature: FP-phase [VEN 11b], FP-group delay [GUP 11, NAI 11], FP-angle [FEN 15], FP-bandwidth [ELA 15], FP-magnitude [VEN 13a], [RAN 15]. These different techniques are compared in terms of the number of bits per resonator, coding density and spectral efficiency in Table 4.1.

The values of coding surface densities given in Table 4.1 are from the corresponding articles. The surface density is often calculated in an "optimistic" way using the dimensions of a single resonator. It does not necessarily reflect the coding capacity of the final tag when several resonators are integrated. When increasing the number of resonators, coupling phenomena appear and can degrade the stated efficiency. In the literature, for the time being, a hybrid coding that uses three different physical quantities does not exist.

Physical quantities used for coding		Types of Resonator		Coding Efficiency (bit/resonator)	Coding Surface Density (bit/cm^2)	Operating Frequency Range (GHz)
Hybrid Coding	Frequency - phase	[VEN 11]	C-resonator	4.6	2.86	2.5– 7.5
	Frequency - angle	[FEN 15]	*Stepped impedance*	3	0.52	1.8 –2.2
	Frequency - bandwidth	[EL-A 15]	Dipole, rect. ring, rect. patch	4	3.6	2 – 5
	Frequency - magnitude	Current study	C-resonator (without ground plane)	3	1.25	2.5 – 6.5
	Frequency - magnitude	Current study	Coupled microstrip dipoles (with ground plane)	6	1.1	3.1 – 7

Table 4.1. *Comparison of different hybrid coding techniques in chipless RFID*

4.1.2. *Magnitude coding method*

In this study, we will examine in detail the possibility of coding information using data related to both the frequency and the magnitude of each resonance peak. The basic principle of frequency coding is used to ensure that each resonator can take several possible frequency values. Beyond that, a similar approach is used for the RCS magnitude, so that the resonators can code several different levels of magnitude. The tag's RCS is therefore quantified both in frequency and in magnitude, as in the case of an analog-digital conversion in signal processing. The basic principle of hybrid magnitude-frequency coding is represented schematically in Figure 4.2.

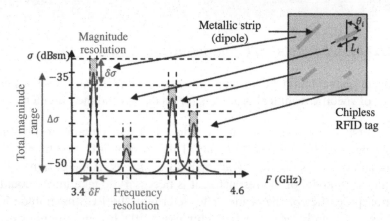

Figure 4.2. *Hybrid frequency-magnitude coding principle. Each resonator (symbolized here by a dipole) is associated with a peak in the spectrum. The information is coded on both the occurrence frequency of the peak as well as its magnitude level. For a color version of this figure, see www.iste.co.uk/rance/rfid.zip*

As in the case for frequency coding, the number of different states coded on the magnitude depends on the total magnitude range $\Delta\sigma$ on which we can vary and measure the presence of the resonator, as well as the magnitude resolution $\delta\sigma$ which characterizes the minimum difference that can be distinguished between two measured peaks. The number of states coded by the magnitude is given by $M = \Delta\sigma/\delta\sigma$. If the geometric parameters used to monitor the magnitude and the frequency are independent, we can consider that the total number of states for a resonator is:

$$N_t = N \cdot M = \frac{\Delta F}{\delta F} \cdot \frac{\Delta\sigma}{\delta\sigma} \qquad\qquad [4.1]$$

This makes it possible to consider high coding capacities without a significant increase in the surface occupied by the tag.

The principle of hybrid coding using magnitude was introduced in [VEN 13a] to enhance the coding density of resonant scatterers with low quality factors, which limits the frequency coding efficiency. This is the case for tags printed on high loss substrates like paper. The magnitude range obtained from this method is relatively low (3 dB) and may not be sufficient for practical applications. There are also a few examples of chipless sensors based on magnitude variations present in the literature [GIR 12a, PRE 11] but the applications are not relevant to our concerns and the studies are limited.

4.1.3. Difficulties related to magnitude coding

The implementation of a magnitude-coded chipless tag poses a certain number of specific technical problems for its design and measurement.

The first question to pose at the time of design pertains to magnitude monitoring in each resonator of the tag. This question is central to magnitude coding and must also be resolved for RCS synthesis. If the design is based on a physical model of the resonator, it is necessary to take into account the radiation properties of the scatterer to determine the RCS magnitude level. This issue is rarely addressed in chipless RFID because monitoring the resonant frequency is sufficient for coding in most cases. The question of the RCS magnitude is therefore neglected despite being of interest to the general framework because it is directly related to the read range.

Magnitude coding also poses practical problems for measurement. We saw in the previous chapter that the magnitude of the response is not an intrinsic parameter of the tag like the resonant frequency or quality coefficient can be. On the contrary, the magnitude depends on the nature of the excitation and the measurement configuration (tag-reader distance, tagged object). The magnitude is also a quantity that is very sensitive to outside perturbations such as the presence of an unknown object in the vicinity of the tag or multiple paths. It is therefore necessary to first study the feasibility of this approach using measurements and in particular to evaluate the magnitude resolution that is sufficient to discern two different peak levels in the presence of noise. It is also necessary to implement compensation techniques to limit the dependence of the magnitude with regard to the measurement setup. We will show that it is useful to add calibration elements within the tag in order to take a differential measurement.

In the remainder of this chapter, we will carry out two complete studies of magnitude-coded tags from design to measurement. The first tag does not have a ground plane, which makes it compatible with printing technologies but also very sensitive to the nature of the object on which it is applied. This tag will be read in co-polarization and consequently the influence of the environment is significant. For this type of measurement, a calibration based on a reference measurement and an empty measurement is necessary. The second tag has a ground plane, which allows for resonances with higher quality coefficients and a better isolation of the tag from the object on which it is applied. This tag will be read in cross-polarization to limit the influence of the environment. In this case, we can remove the calibration step, which is an advantage for applications.

4.2. Tags without ground planes

4.2.1. *Tag design*

The RF Encoding Particle (REP) design approach [PER 14] is used and the elementary particle chosen for the coding is the C-folded dipole that has been studied for a few years now [VEN 11, VEN 12]. The first step in the design is to establish the relations between the geometric parameters of a single scatterer and its electromagnetic signature. For that purpose, two complementary models of scatterer are proposed. To achieve a significant coding capacity, it is necessary to use several scatterers within the same tag. The second step is to integrate these different resonators and account for the coupling effects associated with the chosen configuration. We will show in particular how couplings can enhance the magnitude level of scatterers and potentially increase the total magnitude span available for coding, $\Delta\sigma$.

In [VEN 11], the C-folded dipole has been identified among other single-layer resonators as a good compromise in terms of quality factor ($Q = 65$), RCS level (-30 dB) and frequency range (2.5–7.5 GHz), all while having a reduced size ($L \simeq \lambda/4$). The C-resonator is represented in Figure 4.3(a) with its respective geometric parameters.

A model of this resonator was introduced for the first time in [VEN 11]. The C-folded dipole can be seen as a CPS (coplanar stripline) transmission line in which one of the terminations is connected to a short-circuit (SC) and the other to an open-circuit (OC) as illustrated in Figure 4.3(b). This structure behaves like a quarter-wavelength resonator. The edge effects that appear in the open line modify the resonant frequency, which can be taken into account by introducing a supplementary length ΔL, which is a function of the transverse geometry. As a first approximation, we can consider that ΔL does not depend on the frequency. According to this hypothesis, ΔL can be quickly assessed using an electromagnetic simulation.

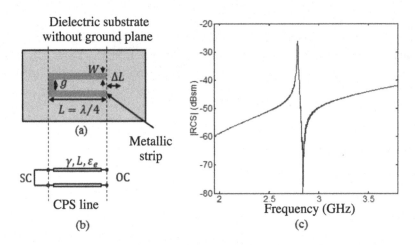

Figure 4.3. *Schematic representation of a C-resonator; a) Geometric parameters associated with the structure; b) Equivalent circuit; c) RCS in the vicinity of the resonant frequency. The C-resonator behaves like a quarter-wavelength resonator. For a color version of this figure, see www.iste.co.uk/rance/rfid.zip*

The resonant frequency is given by:

$$f_r = \frac{c}{4(L+\Delta L)\cdot\sqrt{\varepsilon_e}} \tag{4.2}$$

where the effective permittivity, ε_e, is calculated based on the formulas for an equivalent CPS line [GUP 96] and depends on the gap value g, the width W and the thickness t of the metallic strip as well as the permittivity of the substrate. The equation [4.2] makes it possible to trace the graph of the evolution of the resonant frequency f_r based on the length L for different values of g, (see Figure 4.4). The supplementary length ΔL is obtained by the identification of [4.2] with the resonant frequency obtained in simulation for a given geometric configuration. This type of diagram proves useful for the design because it makes it possible to simply adjust the resonant frequency in relation to the geometric parameters of the resonator.

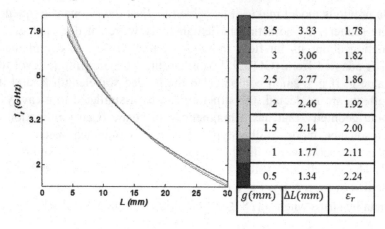

$g(mm)$	$\Delta L(mm)$	ε_r
3.5	3.33	1.78
3	3.06	1.82
2.5	2.77	1.86
2	2.46	1.92
1.5	2.14	2.00
1	1.77	2.11
0.5	1.34	2.24

Figure 4.4. *Calculation of the resonant frequency f_r (logarithmic scale) based on the length of the metallic strips L for different values of gap g. The effective permittivity calculated for the different values of g is indicated in the table on the right. The calculation is carried out for a Fr4 substrate of thickness h = 0.8 mm and permittivity $\varepsilon = 4.6$. The width of the metallic strips is W = 1 mm and the thickness is t = 0.08 mm. For a color version of this figure, see www.iste.co.uk/rance/rfid.zip*

As an example, we can choose to design a C-resonator that resonates at 3.5 GHz (reference case used in the next section). We use a dielectric FR4 substrate of thickness $h = 0.8$ mm and relative permittivity $\varepsilon_r = 4.6$, which corresponds to the diagram in Figure 4.4. We use metallic strips with widths of $W = 1$ mm and a thickness $t = 0.08$ mm, which corresponds to the dimensions used in [VEN 11]. We randomly choose the gap value $g = 0.5$ mm. This value guarantees accuracy for a realization using chemical etching (tolerances in the order of 10 μm for the width of the slot). This value makes it possible to also consider a realization using conductive ink printing. The effective permittivity ε_e of the CPS line is calculated according to [GUP 96] for its geometric dimensions. A single electromagnetic simulation is completed to determine the term ΔL of [4.2]. The equation [4.2] makes it possible to trace the evolution of the resonant frequency based on the length of the metallic strips (Figure 4.4). Once the curve is drawn, we choose the resonant frequency (3.5 GHz) and refer to the length of the corresponding line. In this case, a resonant frequency $f_r = 3.5$ GHz is obtained for a length of $L = 12.7$ mm.

4.2.1.1. Controlling the magnitude – single resonator

The previous model makes it possible to adjust the resonant frequency of a single scatterer by adjusting its length L. However, it does not give any information about its far-field radiation characteristics. A complementary model is adapted from [GOV 95] to relate the RCS magnitude level to the gap value g. If g is small compared to the guided wavelength λ, and if the edge effects are neglected, the structure can be assimilated to a cavity with the same geometry that has a magnetic wall at the open termination. The main resonance mode of this type of cavity is obtained when $L = \lambda/4$; which is in accordance with the result obtained for the transmission line model. The coordinate system used to describe the fields in the cavity is represented in Figure 4.5. For a cavity, the electric field present between the two metallic strips when $-\lambda/8 \leq z \leq \lambda/8$ is given theoretically by:

$$\overrightarrow{E_s} = jI\eta_0 \sin[2\pi(z/\lambda + 1/8)]\,\vec{x} \qquad [4.3]$$

where η_0 is the free-space impedance and I is the total current travelling through the C-folded section (the part where the current circulates according to the $\vec{x}$ axis). To validate the cavity model, this distribution is compared to the fields obtained by the electromagnetic simulation at the resonance. The intensity of the simulated fields is represented graphically in Figure 4.5(a). The theoretical distribution of the normalized field predicted by the model [4.3] is compared to the results obtained through electromagnetic simulation in Figure 4.5(b). We observe good agreement between these results although a difference appears at the open circuit termination for $z > 0.12\,\lambda$. This difference is related to the fringing fields that are taken into account in the simulation as can be observed in Figure 4.5(a).

For small values of g, the cavity model makes it possible to predict the density of the electric current at the surface of the conductor for the first mode:

$$\vec{J}(r) = \begin{cases} \pm I . \cos[2\pi(z/\lambda + 1/8)]\vec{z} \\ I\,\vec{x} \end{cases} \qquad [4.4]$$

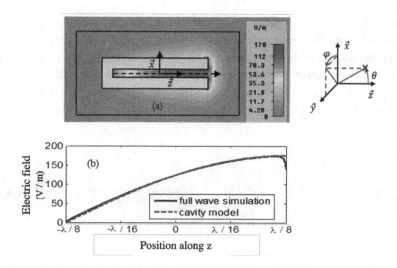

Figure 4.5. *Intensity of the electric field present between the two metallic strips; a)
Coordinate system and results obtained through simulation; b) Comparison between
the simulation (intensity of the field along the dashed line) and the theoretical
prediction [4.3]. For a color version of this figure, see www.iste.co.uk/rance/rfid.zip*

The C-folded section is assumed to have a short electric length
($g < \lambda/50$) so that its behavior is similar to an infinitesimal dipole. This
translates into a constant current distribution. For a longer dipole (for
example, a short dipole: $g < \lambda/10$), the current distribution in the folded part
is not constant, which leads to different results. The currents flowing in
opposite directions in the long metallic strands (oriented following $\vec{z}$)
indicate a low cross-polarization value. The polarization of the antennas
must be aligned with the direction of the current at the folded part ($\vec{x}$) to
ensure the optimal response of the resonator.

The theoretical prediction is compared to the result from the
electromagnetic simulation in Figure 4.6. The simulation shows that the
density of the current is concentrated on the interior faces of the resonator
(slot side). The interior path used to project the density of the current is
represented by black arrows in Figure 4.6(a). The values obtained along the
interior path are compared to the model's prediction and are in good
agreement when a low gap value g is used. For $z > 2$ mm, we find the
expected cosine distribution along the lower metallic strip (Figure 4.6(b)). A
similar result is obtained for the upper metallic strip. A sudden increase in
the current density can be observed for $z < 2$ mm. This is related to the edge

effects (right angle corner) and the thickness of the metallic strips. The calculation path is taken at the middle of the strip thickness but higher density values are observed when closer to the substrate. We can also note that the current density of the short part of the dipole oriented following $\vec{x}$ (Figure 5(c)) is not exactly constant and shows maximums at the corners. A relative variation of 11% is observed compared to the mean value.

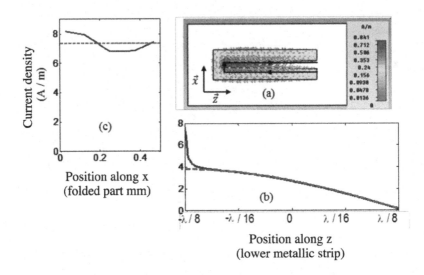

Figure 4.6. *Current density along the interior path at resonance. Comparison between the results predicted by the cavity model and those obtained through simulation (CST Microwave Studio 2012: a) interior path; b) lower arm; c) folded part. For a color version of this figure, see www.iste.co.uk/rance/rfid.zip*

The Fourier transform of the distribution of the current [4.4] makes it possible to calculate the value of the far-field based on the angle of observation [17], so we have:

$$\left\|\overrightarrow{E_{ff}}\right\|^2 = \frac{\eta_0{}^2 (gI)^2}{4\lambda^2 r^2} \operatorname{sinc}^2(\alpha_1) \operatorname{sinc}^2(\alpha_2), \qquad [4.5]$$

with:

$$\alpha_1 = \pi(g/\lambda) \sin\theta \sin\varphi,$$

$$\alpha_2 = \pi(g/\lambda) \sin\theta \cos\varphi,$$

where r is the distance in relation to the point of observation and θ and φ are the elevation and azimuth angles related to the direction of observation as indicated in Figure 4.5(a). The equation [4.5] shows the dependency of the intensity of the field with respect to the length g at resonance. When g is small compared to the wavelength, α_1 and α_2 tend toward zero and the resonator behaves like an isotropic scatterer. In this case, the alignment between the reader and the tag is therefore not a critical parameter at the time of measurement.

The RCS magnitude level for a resonant frequency of 3.5 GHz has been extracted for different values of the gap g from a series of electromagnetic simulations. The result is compared to the prediction of the cavity model [4.5] in Figure 4.7. We observe a marked discrepancy concerning the influence of g when we compare the RCS obtained by simulation and the one obtained from the cavity model. This can be explained by the differences between the idealized structure proposed in [GOV 95] and the real scatterer: the presence of the substrate, the width of the metallic strip comparable to the gap value and a thickness of the strips that is much lower than the gap value. However, a similar qualitative behavior can be observed: the RCS value is mainly sensitive to the value of g. In addition, the distribution of the current at the C-fold depends on the ratio g/λ (relatively constant for an infinitesimal dipole or variable for a short dipole). By analogy to [4.5], we assume that the RCS of the C-folded dipole varies based on g according to the model $\sigma = a \cdot g^b$. The parameters a and b of the model are adjusted to best fit with the maximums at the least squares sense. When g belongs to the infinitesimal dipole region ($g < \lambda/50$, i.e. $g < L/12.5$ because $\lambda \approx 4L$), the values obtained are $a = 0.11\ m^2$ and $b = 0.5$. These values correspond to the dashed blue "infinitesimal dipole" fit line in Figure 4.7 and there is a very good agreement between this new model and the results of the simulation in the region of the infinitesimal dipole (Figure 4.7(a)). When g belongs to the region of the "small dipole" ($g < \lambda/10$, i.e. $g < L/2.5$), the values obtained are $a = 0.0075\ m^2$ and $b = 0.16$. Again, there is a good agreement between the model (dashed green line in Figure 4.7) and the results of the simulation for the small dipole region (Figure 4.7(b)). This shows that by modifying the influence of g in [4.5], it is possible to extend its domain of validity so as to also model C-structures with gaps g that are much greater than those considered in the cavity model.

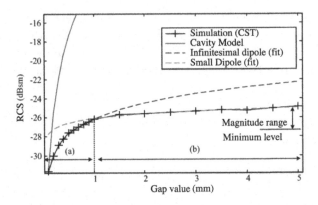

Figure 4.7. *RCS magnitude level at a resonant frequency of 3.5 GHz with respect to the gap value g; a) Infinitesimal dipole region (g < λ/50); b) Small dipole region (g < λ/10). L = 12.7 mm, W = 1 mm. The simulations are carried out on a Fr4 substrate of thickness h = 0.8 mm and permittivity ε = 4.6. For a color version of this figure, see www.iste.co.uk/rance/rfid.zip*

A similar diagram can be obtained for the different values of L using only the three simulations ($g < L/12.5$, $g = L/12.5$, and $g > L/12.5$), which can be used advantageously for design.

The quality factor Q_r (power radiated into space) of the theoretical scatterer has been studied in [GOV 95], when g is small compared to wavelength ($g < \lambda/50$), and is given by:

$$\frac{1}{Q_r} \simeq \left(\frac{g}{\lambda}\right)^2 \tag{4.6}$$

As we showed in the previous chapter (section 3.43), in the presence of losses the total quality factor Q_t is generally lower [VEN 13d] and the overall quality factor can be expressed in relation to the quality factors associated with dielectric loss Q_d and with conductor loss Q_c.

Next, we will consider a variation range for g between 0.5 mm and 5 mm which is compatible with a printing realization process. For the previous example ($L = 12.7$ mm, $W = 1$ mm), the corresponding magnitude range is 2.75 dB (see Figure 4.7) for a single resonator, which is unfortunately not sufficient to perform magnitude coding. In the next section, we will attempt to obtain a larger range using the effects of couplings.

4.2.1.2. Controlling the magnitude – effect of couplings

To increase the coding capacity, several scatterers have been integrated together within the tag. For this study, five C-folded dipoles are arranged in a column in a tag with the dimensions 30 mm × 50 mm, seen in Figure 4.8. This generates strong couplings and the overall response of the tag for each resonant frequency can vary significantly compared to the response of a single resonator.

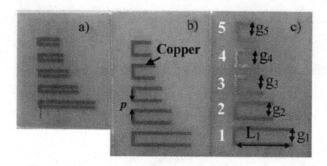

Figure 4.8. *Realized tags. a) Tag 1, b) Tag 3, c) Tag 4. Each tag is composed of 5 C-folded dipoles with the same gap value g. The tag occupies a maximum surface of 30 mm × 50 mm. The scatterers are designed to resonate at 2.5, 3.5, 4.5, and 5.5 GHz. A FR4 substrate of thickness h = 0.8 mm and permittivity ε = 4.6 is used. For a color version of this figure, see www.iste.co.uk/rance/rfid.zip*

mm	Tag1	Tag2	Tag3	Tag4	f_r *(GHz)*
g	0.5	1.5	2.5	3.5	
L_1	18.4	18.9	19.1	19.1	2.5
L_2	12.7	12.7	12.5	12.1	3.5
L_3	9.7	9.4	9.2	8.9	4.5
L_4	7.8	7.4	7.1	6.7	5.5
L_5	6.4	6.1	5.7	5.4	6.5

Table 4.2. *Values of the parameters for the tags realized*

In [VEN 13d], the authors compared the effect of couplings in relation to the configuration for performing frequency coding. They compared the RCS of five resonators with the same gap value and a consistent spacing of $p = 2$ mm, arranged in a line or a column. A frequency offset of maximum

1.5% (30 MHz) compared to the single resonator characteristic was observed. This was interpreted as a supplementary length ΔL_c due to the couplings and it was compensated by adjusting the length of the resonators. For a different code, that is a small variation in the length of each resonator L_i, the couplings remain similar and generally preserve the same value of ΔL_c from one configuration to the next. Therefore, we observe a negligible frequency deviation.

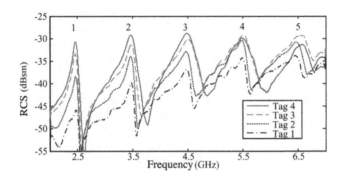

Figure 4.9. *Electromagnetic signature of the realized tags. Each tag shows a different RCS level that can be changed by adjusting the gap value g. For a color version of this figure, see www.iste.co.uk/rance/rfid.zip*

A parametric study has been performed to determine the magnitude sensitivity of the peaks to the gap values in the presence of couplings. The vertical and horizontal configurations have been tested but the vertical configuration shows the greatest variability. For a vertical configuration, a significant variation is observed when the gap value of all the resonators is kept identical. Four tags showing different magnitude values have been realized and measured to illustrate this phenomenon. The results are presented in Figure 4.9, and the measurement setup is detailed later in this section. The values of the parameters are given in Table 4.2. The tags have been realized on a FR4 substrate of permittivity $\varepsilon = 4.6$ and thickness $h = 0.8$ mm. The tags are designed to have the same resonant frequencies at 2.5 GHz, 3.5 GHz, 4.5 GHz, 5.5 GHz and 6.5 GHz.

The variation in the size of the gaps g induces a frequency shift that is compensated by adjusting the value of L (see Table 4.2). In this way, a frequency shift less than 50 MHz is obtained from one tag to another. The spacing between the scatterers is held constant ($p = 2$ mm). The first three peaks (Figure 4.9) show the most significant variations in magnitude (15.2, 10.6 and 8.1 dB respectively for peaks 1, 2 and 3) and can be used for coding in magnitude.

A second study is done when the gap value g_i of a single resonator is modified. The gap value g_2 of the second scatterer is varied from 0.5 mm to 3.5 mm. The other geometric parameters have been kept constant and correspond to those of tag 2 (see Table 4.2). The corresponding magnitude of each peak, is plotted in the graph in Figure 4.10. The magnitude span of the second peak (for which g_2 varies) is $\Delta\sigma = 1.7\ dB$, which is not larger than for a single resonator. It is interesting to note that surrounding scatterers show a sensitivity that is comparable to a variable resonator. This demonstrates that the resonators are strongly coupled.

This strong coupling implies two major difficulties for the design. Firstly, there is no accurate model to describe the effect of couplings for such structures and the design must therefore rely on empirical techniques. However, we note that a coupling model for similar resonators was proposed in [MAC 14], although with limited success. Secondly, due to the coupling, the RCS magnitude level of the peaks cannot be set independently from one resonator to another. To reach distinct magnitude levels with variations in the order of 3dB between two consecutive levels, all of the resonators must have the same gap value. This is a clear limitation for the additional coding capacity offered by magnitude control for this structure.

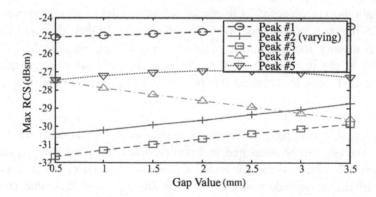

Figure 4.10. *Variation at the RCS magnitude level of the five peaks in relation to g_2. This result shows that there is a strong coupling between the different resonators. For a color version of this figure, see www.iste.co.uk/rance/rfid.zip*

Given the difficulties related to the strong coupling between the resonators, we will limit our design to the case of identical gap values. For the initial tag [VEN 11], which is to say without magnitude coding, six different frequency positions are considered for each resonator, which gives a total coding capacity of $\log_2(6^5) = 13$ bits. If we consider four different magnitude levels, an increase of 2 bits is reached with magnitude coding, which gives a total capacity of 15 bits within a tag with the dimensions 3 cm × 4 cm. It is easy to imagine that designs based on other types of resonators could reach higher coding capacities. In fact, in the hypothetical case where the magnitude of each resonator can be set independently with a comparable magnitude range, this gives a coding capacity of $\ln_2(6^5 \times 4^5) = 23$ bits, which illustrates the potential gain offered by magnitude coding.

We will now examine if it is possible to retrieve the magnitude information in practice. This design, although not optimal, remains relevant to evaluating the feasibility of the approach in similar conditions of application. To do this, it is necessary to determine the minimum magnitude resolution that can be measured. The significant magnitude span of the first few peaks will enable this evaluation on a frequency band from 2.5 to 4.5 GHz.

4.2.2. Measurement results

The RCS magnitude level is known to be particularly sensitive to perturbations. The objective is to assess in practice the minimum variation of the RCS that can be detected in measurement. To do this, different factors that can modify the magnitude of the tag's RCS are examined. The tags designed in the previous section are measured in an anechoic chamber at different distances to estimate the measurement accuracy. The magnitude resolution necessary to distinguish two consecutive levels is deduced from this practical study. The influence of the tagged object is considered as well as the possible presence of an obstacle between the tag and the reader. Finally, the tag will be measured in a real environment. This study is carried out using laboratory tools that make it possible to determine the reference values of the magnitude resolution which are the most favorable (toward which commercial readers will have to move in the near future) [GAR 15].

4.2.2.1. *Effect of the read range*

The theoretical definition of the RCS does not depend on the distance between the tag and the antenna [KNO 04]. However, in practice, the measured quantity is not the RCS but the power reflected at the antenna of the reader, which depends strongly on the measurement setup. The usual way to obtain an exact value of the RCS is to proceed with a calibration measurement. A measurement setup similar to that of [VEN 11] is implemented (Figure 4.11). The measurements are done in the frequency domain with a N52221 network analyzer (VNA) in a bistatic configuration with two vertically polarized antennas. The power emitted by the VNA is 0 dBm in the frequency band from 2–8 GHz. The two horn antennas have a gain of 12 dBi in the frequency band concerned. The spacing between the antennas is $e = 30$ cm. The measured quantities correspond to the S_{21} parameter of the VNA. The tag is placed at a minimum distance of 60 cm from the antennas ensuring far-field conditions for the whole frequency band. An isolation measurement with no tag (S_{iso}) is carried out to characterize the coupling between the transmitting and receiving antennas. A reference measurement (S_{ref}) is also carried out with a rectangular metallic plate whose RCS is known (σ_{ref}). This makes it possible to extract the exact value of the RCS using [3.16].

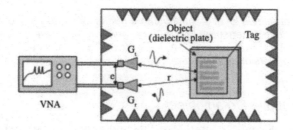

Figure 4.11. *Bistatic measurement configuration. The measurements are realized in an anechoic chamber. For a color version of this figure, see www.iste.co.uk/rance/rfid.zip*

The curves shown in Figure 4.12 correspond to the RCS calculated with [4.16] for the tags 1 to 4 positioned at a distance $r = 60$ cm from the reader. The RCS extraction is also performed on tags positioned at 70 and 80 cm. The differences in the peaks are represented by purple error bars in Figure 4.12. A maximum deviation of 2.1 dB is observed for the first peak of tag 1, which corresponds to the lowest magnitude value (–45 dB), and is therefore more sensitive to noise. In the other cases, the measurement varies by less than 1.4 dB. The magnitude resolution in dB must therefore be higher than 2.1 dB to be capable of distinguishing the magnitude of each peak at a distance of 80 cm. When the distance increases, the attenuation factor related to the propagation in free space increases and the power reflected to the antennas is lower. The signal-to-noise ratio tends to decrease and a higher frequency resolution will certainly be necessary for $r > 80$ cm. One way to improve this type of coding is to consider a variable magnitude resolution based on the magnitude level or even the frequency.

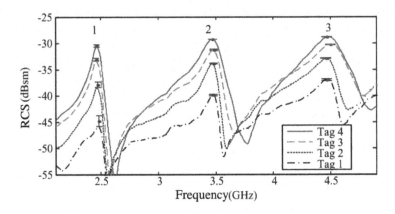

Figure 4.12. *Magnitude level of the first three peaks measured at a distance $r = 60$ cm. The peak levels measured at $r = 70$ cm and $r = 80$ cm are indicated by error bars. A maximum deviation of 2.1 dB is observed for the first peak in tag 1. For a color version of this figure, see www.iste.co.uk/rance/rfid.zip*

4.2.2.2. Tag applied to an object

In practice, the tag is applied on an unknown object that has its own electromagnetic signature. The total backscattered field $\overrightarrow{E_t}(f)$ is the sum of two components: $\overrightarrow{E_t}(f) = \overrightarrow{E_r}(f) + \overrightarrow{E_o}(f)$ where $\overrightarrow{E_r}(f)$ is the response of the tag and $\overrightarrow{E_o}(f)$ is the response of the object. These two components are functions of the frequency. For everyday objects, $\overrightarrow{E_o}(f)$ is generally due to the specular reflection of the incident wave on the object and we therefore get a broadband response, limited in time. Initially, we will consider that there is no interaction between the tag and the object: in other words, that a modification of the object's characteristics modifies only $\overrightarrow{E_o}(f)$ without affecting $\overrightarrow{E_r}(f)$. The effects of an interaction will be examined later. The object contribution considered to be unknown introduces an additional component that can result in an erroneous detection of the identifier. To remove this contribution, it is possible to perform a differential measurement.

Consider two different tags. These tags have been applied on the same object and the corresponding backscattered field is measured. The field backscattered by the ith tag is noted $\vec{E}_{ti}$. We note ΔE_{ij}, the difference between the fields reflected by two different tags:

$$\Delta E_{ij} = \left| \vec{E}_{ti} - \vec{E}_{tj} \right| = \left| \vec{E}_{ri} - \vec{E}_{rj} \right| \qquad [4.7]$$

This differential quantity no longer has a direct dependence on $\overrightarrow{E_o}$ and makes it possible to assess the influence of the interaction between the object and the tag (the variation of the term $\overrightarrow{E_r}(f)$) in practice. For simple objects like thin dielectric plates, it is possible to implement a compensation technique that makes it possible to isolate the term $\overrightarrow{E_r}(f)$ even when a single tag is measured. The magnitude resolution can then be assessed based only on the variations of $\overrightarrow{E_r}(f)$ and not the total response (object +tag) $\overrightarrow{E_t}(f)$.

For tags with no ground planes, the interaction between the tag and the object can significantly modify the response of the tag itself ($\overrightarrow{E_r}$ in the previous equations). We mainly find two effects. Firstly, the object will modify the effective permittivity of the tag. If the permittivity and thickness of the object are known, the effective permittivity ε_r' associated with the resonator can be calculated analytically in a similar way to the CPS line with a multilayer substrate [GUP 96]. By using ε_r' in [4.2] in the place of ε_r we obtain the new resonant frequency f_r' of the resonator:

$$f_r' = \frac{c}{4(L+\Delta L)\sqrt{\varepsilon_{e'}}} \qquad [4.8]$$

Therefore, it is possible to express f_r' as a function of f_r:

$$f_r' = \frac{\sqrt{\varepsilon_e}}{\sqrt{\varepsilon_{e'}}} \cdot f_r \qquad [4.9]$$

From which it appears clearly that the resonant frequency is shifted by a factor of $\sqrt{\varepsilon_e}/\sqrt{\varepsilon_e'}$. This phenomenon has been studied in [VEN 12c] and a compensation technique based on the presence of a reference resonator was proposed. This reference resonator is present in each tag or, in other words, it is not used for coding because it is identical from one tag to another.

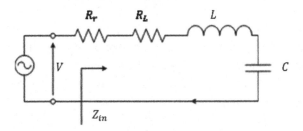

Figure 4.13. *RLC series circuit that is equivalent to a resonant scatterer in the presence of loss*

Secondly, the presence of an object introduces additional dielectric losses that can modify the quality factor of the resonators as expected based on [3.43]. For a resonator, there is a direct relation between the power radiated and the quality factor. For example, we assume that the resonator behaves like a resonant antenna used as a scatterer. As we showed in Chapter 3, by only considering the antenna mode we have the relation:

$$\sqrt{\sigma} = G\lambda/2\sqrt{\pi}$$

where the gain can be expressed as a function of the radiation efficiency e_r and the directivity D of the equivalent antenna:

$$G = e_r \cdot D \qquad\qquad [4.10]$$

Around the resonance, the antenna can be modeled by an equivalent RLC series circuit [BAL 05] as represented in the Figure 4.13 where R_r is the resistance associated with the radiation and R_L is the resistance associated with loss. The radiation efficiency corresponding to the circuit in figure 4.13 is given by:

$$e_r = \frac{R_r}{R_r + R_L} \qquad\qquad [4.11]$$

It is possible to define a quality factor for each loss source. By using the relations of the RLC series circuit presented in the previous chapter we have:

$$Q_r = \frac{\omega_0 L}{R_r} \qquad Q_t = \frac{\omega_0 L}{R_L + R_r}$$

Which allows us to write:

$$G = \frac{Q_t}{Q_r} \cdot D \qquad\qquad\qquad\qquad\qquad [4.12]$$

And finally:

$$\sqrt{\sigma} = \frac{Q_t}{Q_r} \cdot D\lambda / 2\sqrt{\pi} \qquad\qquad\qquad\qquad [4.13]$$

Although this is not a perfectly rigorous representation because the structural mode is neglected, we can clearly see that the presence of the dielectric implies a variation of the RCS magnitude level due to the additional loss.

The presence of an object can also modify the intensity of the couplings that exist between the resonators of a single tag. Modifying couplings between resonators causes other effects like modifications of the re-radiation patterns or a variation in the magnitude of peaks. The absence of an analytical model for couplings makes these effects more difficult to characterize. A sufficient magnitude resolution must be determined in order to account for potential perturbation effects.

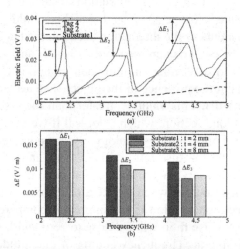

Figure 4.14. *a) Differential measurement of the reflected fields ΔE by tags 2 and 4 applied to the surface of the dielectric $t = 2\,mm$, $\varepsilon_r = 3$; b) Variation of ΔE with respect to the thickness of substrates with the same permittivity ($\varepsilon_r = 3$). For a color version of this figure, see www.iste.co.uk/rance/rfid.zip*

As noted previously, the magnitude resolution required to distinguish two successive magnitude levels will be evaluated in practice starting from the differential quantity ΔE_{ij}. To verify that the differential quantity ΔE_{ij} presents variations that are lower than those of the total response (tag + object), measurements of tags applied on different objects have been performed. The objects are made of dielectric plates with the dimensions 45 mm × 55 mm and of variable thickness $t = 2$ mm, 4 mm, and 8 mm, numbered 1 to 3 respectively. The relative permittivity of the plate is $\varepsilon_r = 3$. For instance, tags 2 and 4 have been successively applied to the surface of the plates and the corresponding backscattered fields are measured. The presence of several resonators at the level of the tag makes it possible to verify the relative invariance of ΔE_{24} on a frequency band of 2.5 GHz to 4.5 GHz and for different magnitude values.

Figure 4.14(a) shows the electric field response backscattered by tags 2 and 4 positioned on substrate 1. The value ΔE_{24}^i indicated by a double arrow in Figure 4.14(a) represents the difference between the re-radiated fields of the two tags for the ith peak. The values of ΔE_{24}^i obtained for the three dielectric plates are compared in Figure 4.14b. The resonant frequency of the peaks shift due to the variation of the thickness of the plates. The value is almost constant for the first peak. A variation of 20% and 30% of the total magnitude is obtained for peaks 2 and 3. This could be due to either inaccurate measurements or the interaction between the tag and the substrate. The relative variation can be expressed in dB (3.1 dB and 1.9 dB for peaks 2 and 3 respectively) to determine an equivalent magnitude resolution in dB that takes into account the presence of the object. The necessary resolution is considered to be 3.5 dB. This value is greater than the resolution of 2.1 dB obtained in the previous section, which is logical if we consider the object as an additional noise. Following this analogy, by considering an object with a higher RCS, the signal-to-noise ratio will decrease and a higher magnitude resolution will be necessary. A limit condition beyond which measurement conditions are difficult occurs when the RCS of the object is equal to that of the tag. If the object contribution is too important compared to the RCS of the tag, the signal-to-noise ratio will be close to zero and it will no longer be possible to measure the tag (frequency coding or magnitude coding). In this case, it is necessary to include the object in the environment for the isolated measurement S_{iso} during the calibration phase. More elaborate detection methods based on the temporal separation between the tag and the object can also be implemented [REZ 15a].

For objects with a low RCS (RCS value comparable to the dielectric plates described in this section), a resolution of 3.5 dB is considered sufficient to discriminate successive magnitude levels. Since the magnitude range of the tags realized is 15.1 dB, four levels of magnitude can be set, which corresponds to an additional 2 bits of coding capacity.

4.2.2.3. *Tag coated by a dielectric*

An interesting case of application occurs when a tagged object is placed inside some packaging, such as a cardboard box or an envelope. In this case, there is no direct line of sight between the reader and the tag. The obstacle attenuates the signal, which can lead to detection errors. Similar to the previous section, a relative measurement can be performed to prevent this type of error. The measurements of tags coated by two different dielectric plates (denoted by 1 and 2) of thickness $t_1 = 0.83$ mm and $t_2 = 0.68$ mm, permittivity of $\varepsilon_1 = 5.1$ and $\varepsilon_2 = 2.7$ and loss tangent of $\delta_1 = 0.18$ and $\delta_2 = 0.02$ have been completed. The distance between the tags and the reader is 60 cm and the dielectric plates are applied directly on the tag in such a way as to also modify the effective permittivity of the CPS line. The measurement results of the backscattered fields are presented in Figure 4.15. Similar behavior to that of a tag applied on a dielectric is observed. The change in permittivity due to the presence of the dielectric in the vicinity of the tag induces a frequency deviation of 450 MHz and 140 MHz for dielectric 1 and 2 respectively. If the dielectric plate does not directly touch the tag, we do not observe a frequency shift. The quality factor of the resonance is lower when the tag is applied behind the dielectric 1 due to higher loss.

A comparison of the differential quantity ΔE_{34} for an uncoated tag and for a tag coated by the dielectrics 1 and 2 is carried out in Table 4.3. Like in the previous section, ΔE_{34} remains nearly constant when the dielectric has a low loss. When the dielectric plate has a higher loss, we can see that ΔE_{34} varies in a similar way for all of the peaks. A factor of 0.5 appears for the dielectric 2 compared to the uncoated case. The resolution of 3.5 dB that was determined in the previous section is sufficient with regard to the variation of ΔE for this configuration. Like in the previous section, the limit case is reached when the RCS of the tag is equal to that of the dielectric plate.

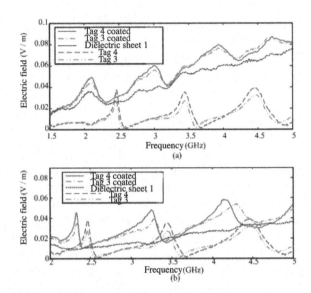

Figure 4.15. *Differential measurement of tags 3 and 4 in an anechoic chamber. a) Re-radiated field for tags 3 and 4 coated by a dielectric plate of thickness $t_1 = 0.83$ mm, permittivity of $\varepsilon_1 = 5.1$, and loss tangent $\delta_1 = 0.18$; b) Re-radiated field for tags 3 and 4 coated by a plate that is $t_2 = 0.68$ mm, $\varepsilon_2 = 2.7$ and $\delta_1 = 0.02$. For a color version of this figure, see www.iste.co.uk/rance/rfid.zip.*

	ΔE_{34}^1 (mV/m) peak n°1	ΔE_{34}^2 (mV/m) peak n°2	ΔE_{34}^3 (mV/m) peak n°3
No sheet	10	8	6
Dielectric sheet n°1	5	4	3
Dielectric sheet n°2	11	9	5

Table 4.3. *Relative measurements of coated tags*

4.2.2.4. *Real environment measurement*

The tags have been measured in an office environment. The distance between the tag and the reader has been decreased to 20 cm given the more significant noise in this type of environment. In order to ensure the far-field condition at this distance, horn antennas used in anechoic chambers have been replaced by Satimos QH2000 antennas of a smaller size. The measurement is realized with a bistatic configuration. The antennas present a gain that varies between 3 dBi and 8 dBi in the frequency band under consideration (2 GHz–5 GHz). A photograph of the measurement configuration is shown in Figure 4.16(a).

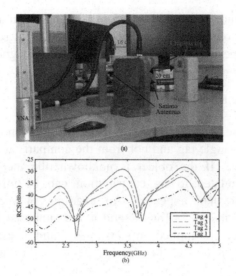

Figure 4.16. *a) Measurement setup in an office environment. The distance between the tag and antennas is 20 cm. Satimo QH2000 antennas have been used; b) Real environment RCS measurement. For a color version of this figure, see www.iste.co.uk/rance/rfid.zip*

Tags 1 to 4 have been measured and their responses are presented in Figure 4.16(b). A rectangular time window has been applied to the measured data to limit the impact of the reflection on the wall visible in Figure 4.16(a) (see section 3.5.2). The magnitude levels of the second and third peaks are similar to the results obtained in the anechoic chamber (Figure 4.15). For instance, considering the second peak at 3.5 GHz of tag 4, an RCS value of −28 dBsm is obtained in a real environment compared to −28.5 dBsm in an anechoic chamber. For all of the tags, the first peak presents a lower magnitude level than the anechoic chamber measurement. This difference can be explained by both the interference of WiFi at 2.4 GHz and by the lower gain of the antennas at this frequency. Nevertheless, the distance between the different levels of magnitude remains similar to the results obtained in the anechoic chamber (2.1 dBsm between the first peaks of tags 3 and 4 in a real environment, compared to 2.3 dBsm in an anechoic chamber). The frequency position of the peaks remains identical for the two configurations. The selectivity of the peaks is lower than in the anechoic chamber, which can be explained by the higher noise level. The measurements realized in a real environment provide results comparable to an anechoic chamber. The main difference is the lower read range that can occur. The magnitude resolution of 3.5 dB that has been assessed based on the measurements obtained in the anechoic chamber is sufficient even for real applications because no critical degradation of the signal is observed.

4.2.3. *Compensation technique*

For practical use, it is necessary to determine the identifier related to the magnitude from a single tag and not from the comparison between two tags as described above. If the object is unknown, the increase in the RCS magnitude level related to its contribution can be misinterpreted as a different identifier than expected. An example of such a scenario is represented in Figure 4.17. Tags 2 and 4 are supposed to code different information. If they are applied on substrates 1 and 4 respectively, we can see that the magnitude of the second peak is the same for the two configurations and it is therefore impossible to correctly recuperate the information from the tags. In addition, a shift of the resonant frequency is observed. A compensation method has been proposed for the frequency coding [VEN 12c] in order to compensate for the shift but without considering the magnitude level. This method can be adapted to find the magnitude code as if the measurement was made in free space (absence of object). The method applies to objects with low RCS values and for which a model of the RCS variation is available. This is the case for thin dielectric substrates. The limit case corresponds, like the previous one, to objects whose RCS is equal to that of the tag.

Like in the previous section, a relative measurement can be realized. However, this time, the calibration element must be included directly in the tag in the form of an additional reference resonator. The magnitude level of the reference scatterer is assumed to be known in the absence of the object. When the tag is applied on an unknown object, the variation in the magnitude of the reference scatterer can be used like a sensor to measure the object's contribution. The reference scatterer must be placed far enough away from the other scatterers to avoid couplings, as illustrated in Figure 4.18. Otherwise, a different identifier can introduce a parasitic magnitude value for the reference and give skewed results. This constraint is not problematic to the extent that, in a tag, the magnitude of each scatterer must be able to be independently adjusted to code information.

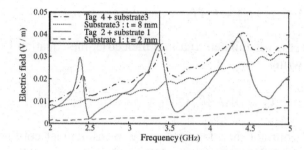

Figure 4.17. *Magnitude level detection error due to the unknown object contribution. The magnitude level of the second peak is similar for tags 2 and 4 when they do not code the same data*

Figure 4.18. *The compensation technique based on a reference resonator. The additional resonator has a known RCS level in the absence of an object. It is used to measure and, subsequently, to remove the object contribution*

As the reference scatterer is a resonator it only provides information at its resonant frequency, not on the entire band under consideration. A model of the evolution of the RCS of the object must be known in order to recover information on the whole band and improve the reliability of this method. If the object can be assimilated to a thin dielectric plate, an approximated formula based on physical optics was obtained in [LE V 85] with very good accuracy:

$$\sqrt{\sigma_s} = \pi\sqrt{4\pi}\,\frac{t\cdot(\varepsilon_r-1)\cdot S}{c^2}\,f^2 \qquad\qquad [4.14]$$

where t is the thickness of the substrate, ε_r is the relative permittivity of the substrate and S is the surface of the plate.

We consider the magnitude level and the resonant frequency of the reference scatterer $\overrightarrow{E_r}(f_0)$ as known in the absence of an object. When the object is present, the measured value $\overrightarrow{E_m}(f_0)$ is modified:

$$\overrightarrow{E_m}(f_0) = \overrightarrow{E_r}(f_0) + \overrightarrow{E_o}(f_0) \tag{4.15}$$

If the object is assimilated to a thin dielectric, the equation [4.18] makes it possible to write:

$$\left|\overrightarrow{E_m}(f_0) - \overrightarrow{E_r}(f_0)\right| = \left|\overrightarrow{E_o}(f_0)\right| = A\pi\sqrt{4\pi}\frac{t(\varepsilon-1)S}{c^2} f_0^2 \tag{4.16}$$

where A is a constant that depends on the measurement configuration and takes into account the attenuation related to propagation in free space. The response of the object for the whole frequency band can therefore be deduced from [4.14] and [4.16]:

$$\overrightarrow{E_o}(f) = \left[\overrightarrow{E_m}(f_0) - \overrightarrow{E_r}(f_0)\right]\left(\frac{f}{f_0}\right)^2 \tag{4.17}$$

Once the response of the substrate is reconstructed on the entire band, it is possible to extract the response of the tag from [4.15].

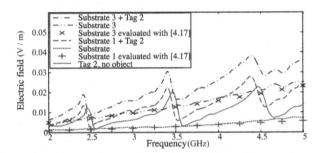

Figure 4.19. *Evaluation of the signature of an unknown substrate (thin dielectric) from the measurement of a reference tag and calculated using [4.17]. The first peak of the "absence of object" configuration is considered to be the reference value. For a color version of this figure, see www.iste.co.uk/rance/rfid.zip*

The compensation technique has been validated in measurement. Tag 2 has been measured for three configurations: in the absence of the object, applied on substrate 1, and applied on substrate 3 (Figure 4.19). As expected, the tag shows magnitude levels that are based on the substrate. A frequency shift of about 100 MHz due to the interaction between the substrate and the tag can be observed. The electromagnetic signatures of the substrates on the whole frequency band are calculated using [4.17]. They are compared with the measurements of the two substrates in Figure 4.19. We observe a good agreement between the reconstructed measurement and the direct measurement.

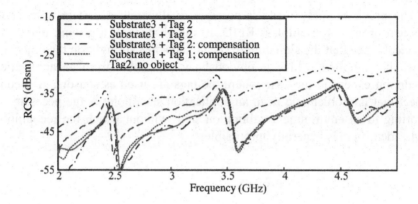

Figure 4.20. *Comparison of the signature of tag 2 obtained using the compensation technique or the measurement in the absence of the object. For a color version of this figure, see www.iste.co.uk/rance/rfid.zip*

After evaluating the response of the substrate, the response of the tag can be found by subtracting the fields. An additional frequency compensation is carried out as in [VEN 12b] to obtain the correct frequency position of the peaks. The compensated signatures are represented in Figure 4.20 and are in good agreement with the measurement in the absence of the object. A maximum deviation of 1.6 dB is observed for the third peak on substrate 3.

4.2.4. *Partial conclusion for tags without ground planes*

The first attempt for coding information at the RCS magnitude level has been successfully achieved. This study, although not optimal from a coding capacity standpoint, makes it possible to identify the theoretical and practical difficulties related to magnitude coding. Four tags with different magnitude levels have been realized and measured. Using the couplings between the individual scatterers, a total magnitude range of 15.2 dB has been obtained. The tags designed make it possible to assess the magnitude resolution necessary to distinguish the different magnitude levels in measurement for a tag without a ground plane. The magnitude resolution has been estimated at 3.5 dB for our study. The problem of the presence of an unknown object has been addressed and a compensation technique has been proposed in the case of a thin dielectric object. This compensation technique has been validated by the measurements.

The previous study made it possible to illustrate a certain number of classical problems in chipless RFID. In the absence of a ground plane, we can clearly see that the signature of the tag depends strongly on the object on which it is applied, which is obviously not desirable from an application standpoint except in the case where the tags are used as sensors. A second aspect that is problematic from an application standpoint is the necessity of resorting to an environment calibration step, without which the recognition of the identifiers is generally impossible.

4.3. Tags with ground planes

The use of a ground plane makes it possible to isolate the tag from the object on which it is applied. This results in an almost negligible tag-object interaction and we can adopt a model of total independence between the response of the tag and that of the object. This eliminates the significant environmental factor during measurement. The second advantage of using a ground plane is that the structure of the tag is similar to microstrip lines or antennas that have been the subject of many studies and are therefore very well documented in the literature. Tags based on microstrip type structures can have higher quality coefficients and consequently higher RCS levels than tags without a ground plane.

Chipless RFID tags generally have RCS levels that are comparable, or sometimes lower, than objects around them. Generally, a prior calibration of the environment is necessary. However, a method that is promising for dissociating the response of the tag from objects in its environment was proposed in [VEN 13c] and presented in the previous chapter. It consists of using depolarizing tags in combination with a cross-polarization reading as was illustrated in the previous chapter. This method allows for a robust readability and a simplified calibration even in the case of a tag applied to a metal object.

This section proposes the implementation of magnitude coding based on the tags proposed in [VEN 13c] that therefore have the advantages presented above.

4.3.1. Tag design

The scatterer used for the coding is represented in Figure 4.21(a). It is composed of n coupled microstrip dipoles with a length L and a width W and separated by a gap g. The metallic strips are positioned on the surface of a substrate with a ground plane. The resonant frequency of the coupled microstrip dipoles is mainly adjusted by the length of the strips L. As a first hypothesis, we can consider the coupled resonators to be half-wave resonators (like a single dipole). For the sake of simplicity, the polarizations of the transmitting and receiving antennas are aligned along the horizontal and vertical axis respectively. The angle between the vertical axis and the dipole axis is indicated by the parameter θ. The dipoles are initially oriented at $\theta = 45°$ to optimize the response in cross-polarization. A study of different values for the angle θ will be presented next. The backscattered E-field obtained through an electromagnetic simulation is represented in Figure 4.21(c) in co-polarization and cross-polarization. The backscattered E-field in co-polarization has a higher magnitude response on a wide band with the presence of dips that are caused by the destructive interference between the field reflected by the ground plane and the resonator. The co-polarization response is therefore dominated by the response of the ground plane or that of the surrounding objects. In a real environment, this leads to a difficult detection if no prior calibration is realized. On the other hand, in cross-polarization, the resonance has a magnitude level that is much higher than that of the ground plane and can be observed easily. For these reasons, we will focus mainly on the response of the tag in cross-polarization.

The REP composed of coupled dipoles is compared to a rectangular microstrip patch used as a scatterer (Figure 4.21(b)). The patch is a well-known structure and there are many theoretical studies in the literature that could be very useful for the design. The lateral dimensions of the two structures are equal so that they cover the same area.

The coupled dipoles and the rectangular patch both resonate at 3.8 GHz. This peak is related to the current path following the direction y (see Figure 4.21(a)). The magnitude level and the quality factor of the resonance are almost identical for both configurations. A second resonance related to the current path following the direction x appears in the case of the rectangular patch. This resonance is eliminated in the case of the coupled dipoles because the gaps that separate them prevent the current from propagating in the direction x. A parasitic antiresonance with a lower magnitude is still observed at 4.4 GHZ because of residual currents in this direction.

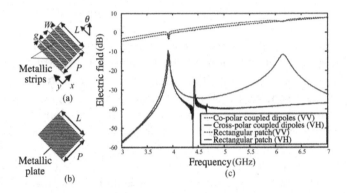

Figure 4.21. *Geometry and signature of the REP used for coding; a) Geometric parameters of the resonator with coupled microstrip dipoles. The local xy coordinate system is attached to the resonator; b) Rectangular patch used as a scatterer; c) Comparison between the signatures of the coupled dipoles and the rectangular patch in co-polarization and cross-polarization. For a color version of this figure, see www.iste.co.uk/rance/rfid.zip*

For a RCS measurement using cross-polarization, the fundamental resonant frequency of a resonator with coupled dipoles is similar to that of a patch with the same dimensions. Many formulas to help with the design are available in the literature for dimensioning a patch antenna, which can be used in an advantageous way for the first step of the design. Electromagnetic simulation softwares are used secondly to perfect the design.

4.3.1.1. *Parametric model*

The resonator with strongly coupled microstrip dipoles is identified by its geometric parameters, which are the input parameters of the model:

$$X = (L, g, W, n, \theta) \qquad [4.18]$$

The response can be described using parameters typical of a resonance, which are considered to be the output of the model:

$$Y = (f_0, A, Q) \qquad [4.19]$$

where f_0 is the resonant frequency, A is the magnitude of the apex of the peak and Q is the quality coefficient associated with the resonance. The frequency response of the resonator in the vicinity of the resonant frequency behaves like a second-order resonator that has a transfer function:

$$S_{vh} = \frac{A}{1+jQ\left(\frac{f}{f_0}-\frac{f_0}{f}\right)}$$ [4.20]

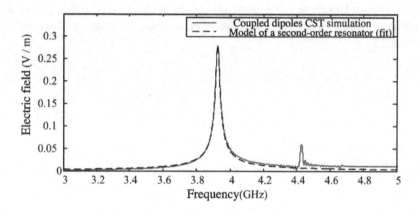

Figure 4.22. *Comparison between the response of a scatterer with coupled dipoles and a second-order resonator whose parameters have been obtained by a least square adjustment based on [4.24]*

For a given set of geometric parameters X, the component S_{vh} is obtained through electromagnetic simulation. The result of the simulation is then fitted with the transfer function [4.20] to identify the parameters that correspond to the output Y of the model. An example of parametric identification is presented in Figure 4.22 for a resonator with the dimensions $L = 20$ mm, $g = 0.5$ mm, $W = 2$ mm, $n = 5$ and $\theta = 45°$, with a substrate of thickness $t = 0.8$ mm and relative permittivity $\varepsilon_r = 3.55$. The response is therefore characterized by: $f_0 = 3.95$ GHz, $A = 0.28$ V/m, and $Q = 120$. This approach provides a quick way to assess the resonance characteristics of a single resonator based on its geometric dimensions. For example, the curve of the resonant frequency with respect to the length of the dipoles is illustrated in Figure 4.23, the other dimensions being the same as previously. The resonator behaves like a half-wave resonator:

$$f_0 = \frac{c}{2(L+2\Delta L)\cdot\sqrt{\varepsilon_e}} \qquad\qquad [4.21]$$

where ε_e is calculated as for a microstrip line and ΔL is a supplementary length of line that takes into account the fringing fields and that can be calculated as for a rectangular patch [BAL 05] or by simulation to also account for the effect of the coupling between the dipoles. In this case, we obtain a supplementary length $\Delta L = 2.5$ mm from a single simulation ($L = 15$ mm). In practice, ΔL depends on the frequency which explains the difference between the model and the results of the simulation presented in Figure 4.23.

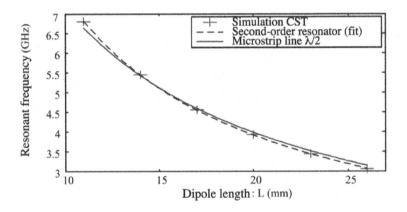

Figure 4.23. *Resonant frequency based on strip length. Comparison between the resonant frequencies obtained through CST full wave simulation, fitting or the model of the half-wave microstrip line [4.21]*

4.3.1.2. *Controlling the magnitude*

Several approaches can be considered to control the magnitude of a tag's signature. A first possibility is to modify the geometry of the tag by varying the value of the different parameters. A second approach consists of duplicating the same resonator several times to reinforce the response at a given frequency. A third option would be to exploit the polarization mismatch between the tag and the antennas. In the upcoming section, these three possible methods will be assessed and compared according to criteria such as the total magnitude range obtained, the induced frequency shift, the selectivity of each resonator, the spatial density of the coding and the ease of implementation.

4.3.1.2.1. Modifying the geometry

A sensitivity study has been conducted to determine the geometric parameters that have the most significant influence on the magnitude of the signature. The reference resonator corresponds to the set of parameters: $n = 5$, $L = 21.8$ mm, $W = 2$ mm, $g = 0.5$ mm. In cases where the value of the parameters is not specified, the dimensions are those of the reference case. The results of the full wave simulations show that the parameter that has the greatest influence on the signature is L. The magnitude level of the apex of the peak as a function of L is represented for different values of n in Figure 4.24. The different curves are all increasing when L increases and have an inflection point for $L = 18$ mm. For the case $n = 5$ and L varying between 10 and 26 mm, an increase of 0.15 V/m (7.2 dB) is obtained. A variation of n induces a magnitude shift in the curve but without notable modification to the general form or the dynamics.

However, the length of the strips L is also the key parameter for determining the resonant frequency of the elementary cell as illustrated in Figure 4.23. The variation range of L corresponds to a shift of f_0 from 3 GHz to 7.5 GHz. Consequently, it is not possible to simply use L to adjust the height of the peaks independently of the frequency because the independence between the resonance and the magnitude of the peak is a necessary condition to implement an efficient hybrid coding.

Other parameters such as n and W contribute substantially to evaluating the magnitude level but with less significant dynamics. As represented in Figure 4.24, for a given value of L ($L = 20$ mm for example), the addition of a supplementary metallic strip (from 5 to 6 strips, for example) increases the magnitude level of the peak by about 0.016 V/m (≈ 1 dB). This variation is also accompanied by a frequency shift, which is approximately 50 MHz here. These two effects become less significant for higher values of n (0.10 V/m and -20 MHz when we move from 6 to 7 metallic strips). The frequency deviation must be compensated by adjusting the value of L, which has an impact on the magnitude level. The frequency accuracy required for coding in frequency position therefore limits the variation range obtained in this way.

The magnitude level of the apex of the peaks in relation to W is represented for different values of n in Figure 4.25. When n varies from 2 to

5, the curves can be considered linear. A maximum increase of $0.047\ V/m$ (2 dB) is obtained for $n = 5$. A frequency deviation of 145 MHz is observed.

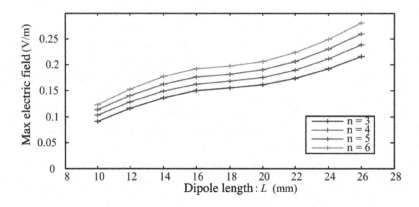

Figure 4.24. *Magnitude A of the peak obtained through simulation (CST) based on the length L of the metallic strips for different numbers of dipoles n. For a color version of this figure, see www.iste.co.uk/rance/rfid.zip*

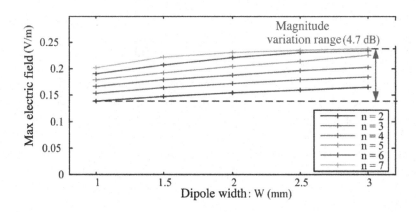

Figure 4.25. *Magnitude A of the peak obtained through simulation (CST) based on the width W of the metallic strips for different numbers of dipoles n. For a color version of this figure, see www.iste.co.uk/rance/rfid.zip*

Similar simulations have been realized to analyze the impact of the parameter g. When g varies between 0.3 and 0.7 mm, a magnitude variation of 0.01 V/m (0.4 dB) with a frequency deviation of only 12 MHz is observed. The parameter g is not very sensitive.

A summary of the results obtained from the sensitivity study is given in Table 4.4. The parameter L, although very sensitive, is the parameter used to adjust the resonant frequency and therefore cannot be used to adjust the magnitude. The total magnitude range achievable from these methods can be evaluated using the simultaneous variations of n and W. As indicated in Figure 4.25 (the magnitude limits are indicated by the dashed lines) a range from 0.139 V/m to 0.238 V/m (4.7 dB) is obtained with this method. The associated frequency shift is 380 MHz. The total width of the structure (P in Figure 4.21(a)) for the case when $n = 7$ and $W = 3$ is equal to 24 mm, which corresponds to twice the value of the reference case. It is interesting to note that this approach requires a large number of electromagnetic simulations and is difficult to transpose to other types of scatterers for chipless tags.

	Range of variation	$\Delta\lvert E^r\rvert$ (V.m^{-1})	Δf_0	Comments
L	10 – 26 mm	0.113 – 0.260 (7.2 dB)	4.5 GHz	Very sensitive parameter, used for frequency coding
n	2 – 7	0.155 – 0.231 (3.5 dB)	240 MHz	Sensitive parameter
W	1 – 3 mm	0.179 – 0.226 (2 dB)	145 MHz	Sensitive parameter
g	0.3 – 0.7 mm	0.199 – 0.208 (0.4 dB)	12 MHz	Not very sensitive parameter

Table 4.4. *Sensitivity study*

4.3.1.2.2. Adding identical resonators

The basic principle of this method is to include the same resonator several times within a tag to increase the intensity of the response at a given frequency. We use m to denote the number of identical resonators that are included in a tag. The resonators are separated by a distance $d = 40$ mm from center to center and positioned in a square as illustrated in Figure 4.26. The resonators are added in an iterative way with m varying from 1 to 4. The total structure can be seen as a network of very closely spaced elements. For a tag illuminated by a plane wave with normal incidence, all of the resonators respond in phase because they are positioned symmetrically in relation to the center of the tag. If we assume that there is no coupling effect between the resonators, the backscattered field is proportional to the number of resonators. The signature obtained through simulation (CST Microwave Studio) for these configurations is represented in Figure 4.26.

The magnitude and the quality coefficient in relation to the number of resonators are represented in Figure 4.27. The magnitude of the peak varies in a linear fashion from 0.3 V/m to 0.6 V/m (6 dB) with m. A frequency shift of 10 MHz is also observed. The resonance region is defined arbitrarily as the frequency band from 3.55 GHz to 3.65 GHz (dashed black lines in Figure 4.26). Outside of the resonance region, the scattering can be assimilated to an optical reflection. We can therefore conclude that the scatterers are not coupled for these frequencies. The magnitude of the response is therefore proportional to the number of resonators (for example at 3.55 GHz, the magnitude levels are 0.044, 0.090, 0.137 and 0.190 V/m when $m = 1$ to 4 respectively). In the resonance region, and in particular for the apex of the peak, the proportionality is no longer observed (0.3, 0.43, 0.1 and 0.60 V/m at the resonant frequency when $m = 1$ to 4, see Figure 4.26).

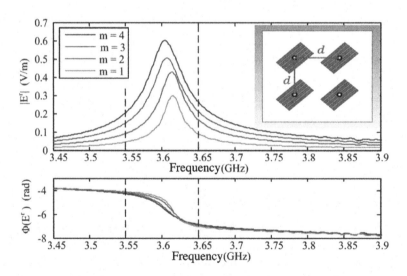

Figure 4.26. *Magnitude and phase of the simulated backscattered field for several identical resonators. For a color version of this figure, see www.iste.co.uk/rance/rfid.zip*

By analogy with the behavior of closely spaced antenna networks, this phenomenon can be explained by the significant couplings that exist between the resonators in that region. The different behavior of the resonators between these two regions (optical and resonant) is directly responsible for the apparent decrease of the quality coefficient (180 to 96 for $m = 1$ to 4) that can be observed in Figure 4.27.

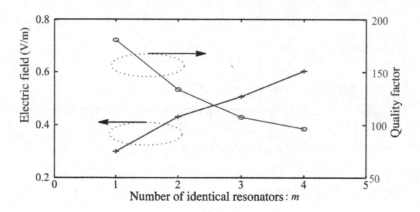

Figure 4.27. *Variation in the magnitude and quality factor based on the number of identical resonators m. For a color version of this figure, see www.iste.co.uk/rance/rfid.zip*

The couplings between two identical resonators have been examined based on a series of electromagnetic simulations. Two different configurations, represented schematically in Figures 4.28(a) and (b), have been considered. The evolution of the magnitude of the peak in relation to the distance between the centers of the resonators is traced in figure 4.28(c) for these two configurations. For the first configuration, the distance d_1 seems to have little effect on the total reflected field. For the variation range considered, the mean value of the magnitude is 0.43 V/m with a variability of about 0.01 V/m around the mean value. To create a compact design, we can therefore position the resonators close to one another in this direction without expecting significant variations in the magnitude level due to the spacing. The magnitude of the total backscattered field is much more sensitive for the second configuration. A variation from 0.49 to 0.57 V/m (1.1 dB) is observed when d_2 varies from 25 mm to 40 mm. At 40 mm, the value obtained is approximately equal to two times that of a single resonator (0.3 V/m). By adjusting the spacing between the resonators, a continuous variation of the magnitude can be obtained, which allows for a more flexible design.

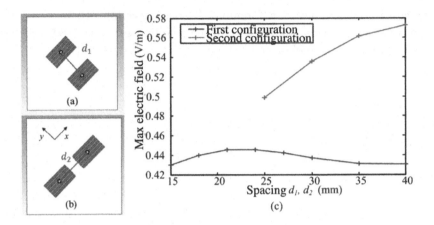

Figure 4.28. *Study of the couplings between two identical resonators; a) First configuration, the resonators are aligned along the y axis (length of strips); b) Second configuration, the resonators are aligned along the x axis (extremities of strips); c) Magnitude of the resonance peak based on the spacing for two identical resonators. For a color version of this figure, see www.iste.co.uk/rance/rfid.zip*

The approach of adding identical resonators makes it possible to actually increase the intensity of the reflected field compared to the case of a single resonator. For the other methods, controlling the magnitude generally causes the intensity of the response to decrease compared to a single optimized resonator. The method can easily be applied to other types of resonators but the performances obtained are difficult to generalize given the coupling effects that depend strongly on the geometry of the single cell. An immediate disadvantage is that the surface occupied by the tag is larger. For a given spacing, if we consider four identical resonators, arranged in a square configuration (Figure 4.26), the total surface of the cell is given by $S = [d + (L + P)/\sqrt{2}]^2$. This must be compared with the surface of a single resonator which is $P \cdot L$. Another disadvantage of this method is the fact that the radiation pattern of the single cell is also modified by what we could define as an array factor. At the resonance, the directivity of the tag increases with the number of resonators, which is a problem for applications of chipless RFID because it makes the alignment between the tag and the antennas more sensitive. Expanding the bandwidth can also pose problems for frequency coding because it limits the number of available frequency windows within the permitted band. Chipless tags based on frequency selective surfaces are mentioned in the literature [COS 13] and can, to a certain extent, be assimilated to the previous method.

4.3.1.2.3. Polarization mismatch

The dipoles are sensitive to electromagnetic waves whose polarization is oriented according to the direction of the strips and they exhibit, in this case, a negligible response in cross-polarization under normal incidence. A simple way to control the magnitude of the backscattered signal is to modify the orientation of the strips (parameter θ) in such a way that the polarization of the antennas is no longer aligned with the direction of the dipoles.

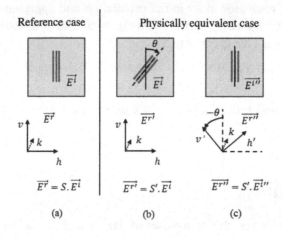

Figure 4.29. *Case study. The antennas are symbolically represented by the polarization vectors v and h; a) Reference case, the dipoles are aligned with the transmitting antenna $\theta = 0$; b) Rotation of the resonator at angle θ with respect to the reference case; c) Rotation of the antennas at angle $-\theta$ with respect to the reference case. The cases (b) and (c) are physically equivalent. For a color version of this figure, see www.iste.co.uk/rance/rfid.zip*

Three different case studies have been examined. They are designated by (a), (b) and (c) in Figure 4.29. We assume that the transmitting antenna and the receiving antenna have vertical and horizontal polarizations respectively. The antennas are symbolically represented by the polarization vectors h and v in Figure 4.29. $\overrightarrow{E^i}$ is the incident field at the tag and its polarization, not represented in the figure, is the same as for the transmitting antenna. For the reference case (a), the resonator is positioned facing the antennas with a reference angle $\theta = 0$. The scattering matrix in the reference case can be obtained using [4.22] and is noted as S:

$$[E^r] = [S] \cdot [E^i] \tag{4.22}$$

We now apply a rotation of angle θ to the resonator (Figure 4.29(b)). A new scattering matrix S' is obtained for this configuration:

$$[E^{r'}] = [S'] \cdot [E^i] \tag{4.23}$$

Physically, applying a rotation of angle θ to the resonator is equivalent to maintaining the resonator in its initial orientation and applying a rotation of angle $-\theta$ to the antennas (Figure 4.29(c)), such that the scattering matrix associated with case (c) is equal to S':

$$[E^{r''}] = [S'] \cdot [E^{i''}]. \tag{4.24}$$

The electric and reflected fields for case (c) are related to those of case (a) by:

$$[E^i] = [\Omega]^T \cdot [E^{i''}]$$

$$[E^r] = [\Omega]^T \cdot [E^{r''}] \tag{4.25}$$

where $[\Omega]^T$ indicates the transpose of $[\Omega]$ which is a transfer matrix corresponding to a rotation of angle θ:

$$[\Omega] = \begin{bmatrix} \cos\theta & -\sin\theta \\ \sin\theta & \cos\theta \end{bmatrix} \tag{4.26}$$

By injecting [4.25] in [4.22] and noting that $[\Omega]$ is a unitary matrix, we find:

$$[E^{r''}] = [\Omega] \cdot [S] \cdot [\Omega]^T \cdot [E^{i''}] \tag{4.27}$$

Through identification with the equation [4.24], this becomes:

$$[S'] = [\Omega] \cdot [S] \cdot [\Omega]^T \tag{4.28}$$

This equation makes it possible to predict the evolution of the scattering matrix when we change the orientation of the tag. From this formulation, it is possible to see that angle θ can be used to modify the measured magnitude of the backscattered field. Formula [4.28] is general and consequently valid for any scatterer. The method is therefore simple to transpose to other types of resonators.

If we consider the case of the coupled dipoles, the scattering matrix takes a very simple form for the reference angle $\theta = 0$ (metallic strips oriented along the vertical axis):

$$\begin{bmatrix} E_h^r \\ E_v^r \end{bmatrix} = \begin{bmatrix} 0 & 0 \\ 0 & S_{vv} \end{bmatrix} \cdot \begin{bmatrix} E_h^i \\ E_v^i \end{bmatrix} \qquad [4.29]$$

Mathematically, the values of the diagonal elements (S_{vv} and 0) are the eigenvalues of the function represented by the scattering matrix. For an isolated target, these quantities are independent to the polarization, which makes them particularly interesting for the identification of targets [WIE 98]. If we express the quantity $S_{vh}(\theta)$ (component related to cross-polarization) in relation to the reference case, we obtain a particularly simple expression:

$$|S_{vh}(\theta)| = |\cos \theta \cdot \sin \theta \cdot S_{vv}|. \qquad [4.30]$$

Based on [4.30], it appears clear that the maximum value that can be read in cross-polarization is obtained when the angle $\theta = 45°$ and corresponds to $S_{vv}/2$.

Equation [4.30] shows that it is possible to control the magnitude of the signature by varying the parameter θ regardless of the frequency. It is important to note that this coding method implies that the orientation of the tag in relation to the reader is known. We assume that this is the case for what follows. The measurement results for the reference resonator ($n = 5$, $L = 21.8$ mm, $W = 2$ mm, $g = 0.5$ mm and $f_r = 3.61$ GHz) are presented in figure 4.30 for different angles θ. The response of the tag has been measured in an anechoic chamber with an N5222A network analyzer at a distance $r = 15$ cm from the antennas. The only calibration that occurred is an empty measurement (no tag or object) intended to characterize the isolation between the antennas.

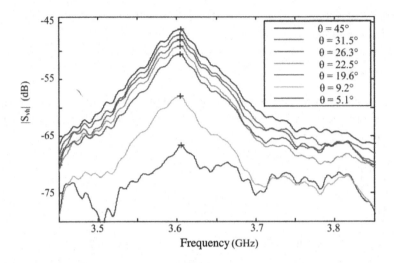

Figure 4.30. *Response of a single resonator ($S_{vh}(\theta)$) for different values of θ. For a color version of this figure, see www.iste.co.uk/rance/rfid.zip*

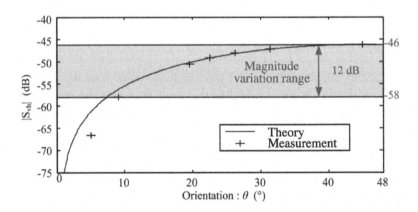

Figure 4.31. *Maximum magnitude as a function of the orientation θ. Comparison between the value calculated according to [4.30] and the measurement results. For a color version of this figure, see www.iste.co.uk/rance/rfid.zip*

A comparison between the theoretical results calculated using [4.30] and the measurement results is presented in Figure 4.31. The crosses represent the locations of the maximums (magnitude of the response at 3.61 GHz). A good agreement is observed between the theory and the measurement. The errors observed for angles less than 10° are due to the very low values of the field (comparable to ambient noise), as well as at the sensitivity that increases when θ approaches zero. In order to avoid these errors, the angular range is limited to values that vary between 8° and 45°, which provides a magnitude range of 12 dB available for coding. It is important to note that with this method for controlling the magnitude we do not observe any frequency shift or even a degradation of the quality factor.

A comparison of the different methods to control the magnitude is summarized in Table 4.5. Controlling the magnitude based on the orientation of the scatterer gives a magnitude range that is the largest and simplest to implement.

Method used	Parameters	Addition	Polarization
Total magnitude range	4.7 dB	6 dB	12 dB
Frequency shift	380 MHz	10 MHz	-
Variation in the quality factor	-	175 - 100	-
Surface occupied	$2.P.L$	$\left[d + (L+P)/\sqrt{2}\right]^2$	$\pi.(L/2)^2$
Adaptability to other types of resonators	No	Yes	Yes
Analytical formula	No	Yes	Yes
Notes	Large number of simulations	Complicated treatment of couplings	Orientation in relation to reader known

Table 4.5. *Comparison of different methods to control magnitude*

The magnitude coding method can only be successfully performed if the lowest resonance peaks of the magnitudes are measurable. The result is a reduction of the maximum read range compared to the case where only the maximum magnitude peaks must be detected (which is the case when using only frequency position coding). Limiting the read range is inherent to the magnitude coding approach and therefore does not depend on the method used to control the magnitude. The read range can be assessed using the radar range equation, see [3.13]. Therefore, we have:

$$R_{max} = \left[\frac{P_t G^2 \lambda^2 \sigma}{(4\pi)^3 P_{min}}\right]^{1/4} \qquad\qquad [3.14]$$

We consider that the RCS of the tag is reduced by 12 dB, which corresponds to the range of magnitude obtained using the polarization method. The RCS of the lowest peak is given by $\sigma' = \sigma/\alpha$ with $\alpha = 10^{12/10}$. The maximum read range R_{max}' corresponding to σ' can be calculated using [4.31] and is related to R_{max} by:

$$R'_{max} = \left[\frac{1}{\alpha} \cdot \frac{P_t G^2 \lambda^2 \sigma}{(4\pi)^3 P_{min}}\right]^{1/4} = \frac{R_{max}}{\alpha^{1/4}} \qquad [4.31]$$

For a total magnitude range of 12 dB, the read range is divided by two compared to an exclusively frequency-based coding. It is interesting to note that for a given system, the "reduction coefficient" does not depend on the type of antenna used but only on the power level that is effectively dedicated to magnitude coding.

4.3.2. Measurement results

The study in the previous section showed that the most efficient approach for controlling the magnitude of a coding cell is to rotate the dipoles in the ground plane. A photograph of the tag realized following this principle is given in Figure 4.32. The tag is composed of four resonators with coupled microstrip dipoles. Each resonator is composed of five metallic strips with a width $W = 2$ mm and separated by a gap $g = 0.5$ mm. The length of the strips and the corresponding resonant frequency for each of the resonators is given in Table 4.6. For a coding in frequency position, each scatterer can be assigned to different frequency windows. The resonators have therefore not been distributed uniformly across the frequency band. Table 4.6 indicates a decrease in the quality factor Q for the highest frequencies, which is related to the fact that only parameter L is modified from one resonator to another. Higher values of Q can be obtained if a scale factor k is applied to the entire structure (keeping the identical proportions). The resonators have been realized on a dielectric Rogers (RO4003C) substrate of permittivity $\varepsilon_r = 3.55$ and thickness $t = 0.8$ mm. The resonators have been placed on the surface of a metallic plate. Each resonator is realized individually in order to be able to modify its orientation easily. The uncertainty on angle θ related to the manual positioning of each resonator has been assessed at approximately $1°$ based on a series of 20 identical measurements. This uncertainty on angle θ can translate into an uncertainty pertaining to the magnitude of the response by deriving the expression [4.30]:

$$dS_{vh}(\theta) = d\theta \cdot cos(2\theta) \cdot S_{hh(\theta=0)} \qquad\qquad [4.32]$$

#	L (mm)	f_r (GHz)	λ_0 (cm)	Q
1	23	3.4	8.8	195
2	21.8	3.6	8.3	173
3	17.9	4.3	6.7	121
4	16.8	4.6	6.5	93

Table 4.6. *Dimensions and characteristic quantities of the tags realized*

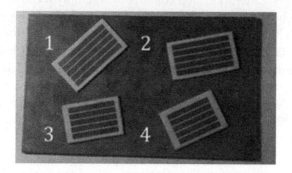

Figure 4.32. *Tag realized composed of four resonators with coupled dipoles. Rogers (RO4003C) substrate, $\varepsilon_r = 3.55$, thickness, $t = 0.8\ mm$*

From [4.32], we can see that an error in the orientation has a greater relative influence when θ is near zero. This uncertainty is definitely less significant for tags realized in a single piece.

As will be demonstrated next, a magnitude resolution of 1.5 dB is sufficient to ensure the correct detection of the magnitude level in practice for a read range up to 27 cm. Compared to the study of the tag without a ground plane, this more reliable magnitude resolution can be explained by the fact that tags with a ground plane present almost no interaction with objects. The total magnitude range of 12 dB considered for this approach makes it possible to define eight different magnitude states for each resonator. The code associated with a tag is composed of four digits $X_1X_2X_3X_4$, where X_i indicates the state taken by the resonator number i and

X_i can vary between 0 and 7. The value 0 identifies the highest magnitude level and 7 identifies the lowest magnitude level.

The measurements have been carried out in an anechoic chamber with a N5222A network analyzer in a monostatic configuration and for a measurement in cross-polarization. The power emitted by the VNA is 0 dBm in the frequency band from 2 GHz to 8 GHz. A Satimo QH2000 antenna with a lateral dimension of 10.5 cm has been used. This antenna has a double polarization, and its cross-polarization isolation is higher than 30 dB. The antenna has a gain that varies from 3 dBi to 8 dBi on the frequency band 2–8 GHz. The only calibration step consists of an empty room measurement to identify the coupling between the two polarizations of the antenna. The measurements have been realized at a distance $R = 30$ cm from the tag. For this distance, the tag is not in the far-field of the antenna but only in the radiating near-field. In this region, the radiation pattern of the antenna is degraded and depends on the read range [YAG 86] but these measurements remain possible, in particular if the tag is placed in the main lobe of the antenna.

4.3.2.1. Contribution of the ground plane

The rectangular metallic plate that serves as a ground plane to the structure is aligned with the polarization axes of the antennas (horizontal and vertical). Electromagnetic simulations of only the plate (without a resonator present) have been carried out. Under normal incidence, the response of the plate in cross-polarization is almost zero (-120 dB in simulation). In practice, however, the plate is not perfectly aligned with the antennas and its response contributes to the tag's signature. The influence of the plate can be observed in Figure 4.33. The response of the reference resonator ($n = 5$, $L = 21.8$ mm, $W = 2$ mm, $g = 0.5$ mm and $f_r = 3.61$ GHz) oriented at the angle $\theta = 45°$ is measured in the vicinity of the resonance. We also measure the plate itself. The response of the resonator minus that of the plate is also represented and is a classic resonance figure. The total backscattered field is the complex sum of the fields scattered by the plate and by the resonator. This involves constructive or destructive interferences depending on the relative phases of the signals. The magnitude level of the plate can be considered to be a noise threshold limit that defines the lowest detection level of the signal. It is a significant limitation on the magnitude range available for coding. The contribution of the plate can be seen as a systematic additive error.

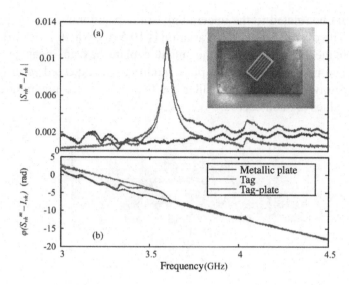

Figure 4.33. *a) Magnitude and b) phase of the field backscattered by the reference resonator (n = 5, L = 21.8 mm, W = 2 mm, g = 0.5 mm and f_r = 3.61 GHz) oriented at the angle θ = 45°. The total field is the complex sum of the fields backscattered by the resonator and the metallic plate. For a color version of this figure, see www.iste.co.uk/rance/rfid.zip*

A series of measurements have been carried out to validate the method of controlling the RCS of the resonator when it is included in a tag with other resonators, that is, in the presence of couplings. Resonator number 3 is rotated so that the magnitude is modified by the theoretical magnitude steps of 1.5 dB [4.30]. The other resonators remain unchanged in the position corresponding to the maximum magnitude (θ = 45°). The measurement results are represented in Figure 4.34(a). A maximum variation of 0.4 dB is observed on the peak corresponding to the resonator number 4 (stationary). This shows that the couplings induced by the variable resonator have a low influence compared to the overall dynamics. Resonator number 3 shows a variation range of 7 dB which is less than the 12 dB expected based on the theoretical study carried out previously (single resonator). This can be explained by the supplementary contribution of the metallic plate that limits the variations for low magnitude levels and also by the influence of couplings. The same measurement results are represented in Figure 4.34(b) after subtracting the contribution related to the metallic plate. In this case, the highest magnitude peaks decrease according to a magnitude step of 1.5 dB as predicted by the theory. For lower peaks the decrease is less regular, which could be related to the uncertainty of the positioning that is greater in this case. The overall magnitude variation is 10.5 dB, which is 3.5 dB greater than when the contribution of the metallic plate has not been taken into account. The contribution of the plate modifies the expected ratio between two consecutive levels, in particular for the case where magnitude levels with low values are considered.

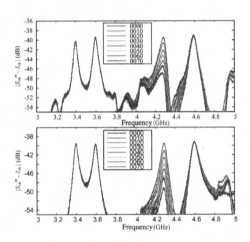

Figure 4.34. *Magnitude of the response measured for different angles of resonator 3 of the tag presented in Figure 3.32. Resonators 1, 2 and 4 are oriented at 45°. The angle of resonator 3 is variable. The angles are determined using [4.30] to have a theoretical magnitude step of 1.5 dB; a) Direct measurement; b) Measurement after subtracting the contribution of the metallic plate. For a color version of this figure, see www.iste.co.uk/rance/rfid.zip*

4.3.2.2. *Effect of the read range*

For traditional measuring equipment, the system is able to measure power that is transmitted P_t and received P_r (phase and magnitude) at the terminations of the VNA. In the absence of calibration, the quantity measured is the transmission parameter of the network analyzer $S_{21}{}^m$. If we assume an ideal case where the measurement does not introduce any errors and the transmitting and reflecting antennas have a vertical and horizontal polarization respectively, S^m is related to the polarimetric RCS through the radar range equation as it was shown in Chapter 3:

$$|S_{21}{}^m| = \sqrt{\frac{P_r}{P_t}} = \frac{G(f)\cdot\lambda}{\left(2\sqrt{\pi}\right)^3\cdot R^2}\sqrt{\sigma_{vh}} \qquad [3.15]$$

where $G(f)$ is the gain of the antenna which is variable on the frequency band and λ is the wavelength in free space. The equation [3.15] clearly shows that $S_{12}{}^{m}$ depends on the read range R as well as the characteristics of the antenna. In the absence of calibration, an unknown read range can therefore skew the detection of the magnitude level of the tag.

For a given measurement, the ratio between two peaks (at resonant frequencies of f_1 and f_2, and by noting λ_1 and λ_2 as the corresponding wavelengths) is given by:

$$\frac{S_{21}{}^{m}(f_2)}{S_{21}{}^{m}(f_1)} = \frac{\lambda_2 \cdot G(f_2)}{\lambda_1 \cdot G(f_1)} \cdot \frac{\sqrt{\sigma(f_2)}}{\sqrt{\sigma(f_1)}} \qquad [4.33]$$

In theory, this does not depend on the read range. A practical way to address the problem of erroneous magnitude detection due to not knowing the tag-reader distance is therefore to consider one of the peaks as a reference element. All of the other peaks are then expressed relative to the value of the measured magnitude corresponding to this reference element. In a certain way, this comes back to including a calibration element directly within the tag. The magnitude resolution must be determined according to a logarithmic scale (dB) to depend only on the ratio between two signals and therefore be independent of the distance.

The same tag has been measured at different distances to verify that the difference between two peaks (logarithmic representation) is almost independent of the distance. The measurement results are represented in Figure 4.35. The first peak is taken as the reference value and is therefore not used to code information. The separation between the ith peak and the first is noted as $\Delta_{1i}S^m$. The differential measurement $\Delta_{12}S^m$ is indicated by a black arrow in Figure 4.35. The separation between the peaks for different read ranges is shown in Table 4.7. Contrary to expectations, over the entire distance range considered (14 to 50 cm), the differential measurements vary significantly with distance. In addition, the magnitude levels of the first and second peaks are difficult to distinguish above 50 cm. This is related to the fact that the reflected power is lower at this distance and the result is therefore more influenced by measurement errors or noise in the environment. It is also likely that this is a consequence of the near-field measurement for smaller distances. The magnitude resolution (in dB) is evaluated at a distance of 27 cm. We define it as twice the maximum deviation observed at this range of distances. This gives a magnitude resolution of 1.5 dB. For greater distances, the signal-to-noise ratio is lower which imposes a higher magnitude resolution.

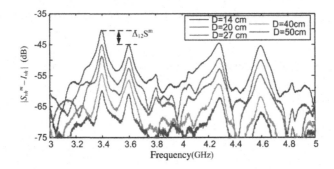

Figure 4.35. *Signature of a tag coding the identifier 0333 measured for different read ranges D (logarithmic scale magnitude). For a color version of this figure, see www.iste.co.uk/rance/rfid.zip*

D (cm) :	14	20	27	40	50	Maximum deviation $(D \leq 27\ cm)$
$\Delta_{12}S^m$ (dB)	4.6	4.0	3.9	2.3	1.1	0.7
$\Delta_{13}S^m$ (dB)	4.2	4.5	4.1	3.5	2.4	0.3
$\Delta_{14}S^m$ (dB)	5.1	5.3	5.8	7.6	7.2	0.7

Table 4.7. *Separation between the peaks for different read ranges*

4.3.2.3. *Measuring random identifiers*

An identifier $X_1X_2X_3X_4$ is chosen randomly and the corresponding tag is realized using [4.30]. The tag is then measured to see if it is possible to decode the correct value from the measurement result. The first resonator is used as a reference element, so it is not used for coding information. The magnitude levels of resonators 2, 3 and 4 are read relative to this reference element. The measurements are realized in an office environment at a distance $D = 27$ cm from the antennas. A Hamming time-domain window is applied to the data acquired by the VNA. The lower limit of the Hamming window is taken at 8 ns to eliminate the component of the response related to the reflection on the metal plate. The upper limit of the window is taken at 20 ns, which corresponds to the extinction of resonant modes and makes it possible to limit the impact of possible reflections of the pulse on the walls.

An overall decrease of 6 dB is observed compared to a similar measurement realized in an anechoic chamber which can be explained by the power of the signal filtered by the time windowing. The measurement results are represented in Figure 4.36. The relative levels are in good agreement from one tag to another but are not those expected based on the theory, which can be explained simply by the fact that the dependences in relation to the wavelength and the antenna gain [4.33] have not been taken into account. The different magnitude levels are easily discernable, however. In the second peak, we observe a lower magnitude step than in theory, which can be explained by the interference that occurs between the first and second peak. These measurements show that a difference in frequency greater than 200 MHz must be taken between two consecutive peaks to limit interferences. A practical solution to avoid detecting erroneous levels consists of considering a variable magnitude resolution that accounts for the characteristics in the measurement system. This means that rather than considering a resolution based on the difference between the peaks of the magnitude of the reflected field in simulation, these terms should be balanced by the coefficient $\lambda_i \cdot G(f_i)$ [4.33] that reflects the effect of the measurement.

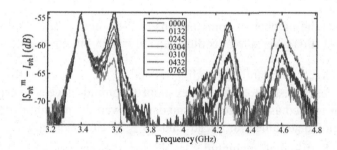

Figure 4.36. *Measurement results for different random identifiers. The first resonator acts as a reference element that is not used for coding. For a color version of this figure, see www.iste.co.uk/rance/rfid.zip*

4.3.3. Coding capacity

Firstly, for the purpose of comparison, calculating the coding capacity is carried out using only the frequency position coding. The number of resonators is indicated by n. In the present case, four resonators are considered. Based on the selectivity that can be reached by the resonators, a realistic frequency resolution of $\delta f = 100$ MHz can be considered. For magnitude coding, we can preserve this resolution to determine the number of frequency windows but we must leave a gap of at least two windows between each resonator. The four resonant peaks are distributed on the operational frequency band that ranges from 3.1 GHz to 7 GHz. Theoretically, the upper limit of the band could reach 9.3 GHz, which corresponds to the second resonance of the coupled dipoles $(3 \cdot \lambda/2)$. In practice, however, according to [VEN 12c] we see that the coupled dipoles are difficult to measure after 7 GHz given the decrease in their quality factor. The most pessimistic perspective is therefore to fix the upper limit of the operating band at 7 GHz. This gives a frequency range of $\Delta F = 3.9$ GHz. The number of corresponding frequency windows is given by: $\Delta F / \delta f = 39$.

From this, the two security frequency windows must be removed. This therefore virtually gives a number $N = \Delta F / df - 2 \times n = 31$ of possible states related to the frequency for each resonator. By considering that there could be a number $k = 1$ to 4 resonators actively present within the tag, the total number of identifiers obtained by considering only frequency coding is therefore:

$$C_F = \sum_{k=1}^{n} C_N^k \qquad [4.34]$$

where C_N^k is the number of combinations of k elements out of N. For $n = 4$ and $N = 31$, this gives a total coding capacity of 15 bits for the tag.

In addition to conventional coding related to the position of the peaks in frequency, each resonator also codes eight magnitude levels, excepting the first which is used as a calibration standard. Given the magnitude range and the resolution determined for a read range D $\leq$ 27 cm, each scatterer can code $M = \frac{\Delta \sigma}{\delta \sigma} = 8$ different magnitude levels for the present case. For every identifier coded in frequency, there are $M^{(k-1)}$ possible identifiers related to the magnitude. For hybrid magnitude-frequency coding, the total number of identifiers is given by:

$$C_{AF} = \sum_{k=1}^{n} C_N^k \cdot M^{(k-1)} \qquad [4.35]$$

For $M = 8$, a total coding capacity of 24 bits is calculated using [4.35]. An additional coding of 3 bits per resonator (the first being used as a reference) is obtained thanks to the magnitude coding approach.

In the example presented, the worst case approach is systematically considered to determine the coding capacity. A more optimistic scenario can also be assessed. A frequency resolution of 50 MHz and a bandwidth that ranges from 3.1 GHz to 9.3 GHz can be considered (130 windows). In addition, the magnitude shift related to the read range *a priori* unknown in the absence of calibration has been doubled in the previous calculation to set the magnitude resolution. By eliminating this factor 2, $M = 16$ levels of magnitude are obtained. If $n = 8$ resonators are integrated into the tag as in [VEN 13c] instead of 4, which is possible for a tag with similar dimensions to a credit card, a total coding capacity of $40.6 + 32 = 72.6$ bits is obtained. For this optimistic case an additional coding of 4 bits per resonator is obtained.

More generally, the supplementary coding capacity provided by magnitude coding can be represented using [4.34] and [4.35]. The coding capacity thus obtained is represented in Figure 4.37 in relation to the number of resonators n for a variable number of magnitude levels M. The hypothetical case of 50 frequency windows is considered. We can see that the magnitude coding method makes it possible to significantly increase the coding capacity even for a relatively low number of resonators and few magnitude levels.

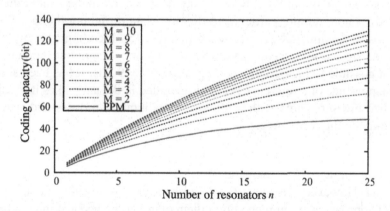

Figure 4.37. *Comparison between the coding capacities associated with only frequency position coding and hybrid magnitude-frequency coding based on the number of resonators n and for different numbers of magnitude states. The calculation is carried out in the context of a pessimistic scenario with 50 frequency windows. PPM (pulse position modulation) indicates the value obtained for frequency coding only. For a color version of this figure, see www.iste.co.uk/rance/rfid.zip*

Table 4.8 provides the number of resonators necessary to reach the objectives of 96 and 128 bits for 8 and 16 magnitude levels. The calculation is done using [4.35] by considering the optimistic case of 130 frequency windows.

	Number of resonators necessary	
	8 magnitude levels	16 magnitude levels
Objective: 96 bits	13 resonators	11 resonators
Objective: 128 bits	19 resonators	16 resonators

Table 4.8. *Number of resonators necessary as a function of the magnitude levels for 96 and 128 bits (130 windows)*

4.3.4. *Partial conclusion – tags with ground planes*

A complete study has been realized to assess the feasibility and the potential gain offered by magnitude coding in the context of chipless RFID. From a comparative study, the simplest way to control the magnitude has been determined. This approach, based on the polarization mismatch between the tag and the antennas, is easy to implement in practice because it only consists of modifying the orientation of the resonators individually, which does not add any particular difficulty when designing the tag. Based on a mostly pessimistic scenario, we estimate a 15-bit coding capacity for a tag made up of four frequency-coded resonators. Magnitude coding has made it possible to increase the coding capacity by 9 bits. It has been shown that the presence of a reference scatterer included in the tag makes it possible to realize measurements with a minimalist calibration process and without depending on the read range.

One limiting aspect that has not been addressed is the influence of the absence of knowledge beforehand of the orientation of the tag in relation to the reader. However, it is theoretically possible to find this information using indirect approaches by implementing them either at the level of the reader or at the level of the tag. For example, an interesting approach consists of comparing the response of two reference resonators that have different, known orientations.

4.4. General conclusion

The method of magnitude coding has been illustrated and evaluated using two examples of different tags. It has been shown that the tag without a ground plane is particularly sensitive to its environment and in particular to the object on which it is applied. For this type of tag, a calibration is necessary for measurement. The quality factors obtained are also relatively low. These different comments are reflected at the coding level by a greater magnitude resolution (3.5 dB in the case of the C-resonator) which limits the number of magnitude states that can be considered for coding information. Tags with a ground plane have the advantage of being isolated from the object on which they are applied. When they are read in cross-polarization, a simplified calibration process can be considered which presents a non-negligible practical interest. A magnitude resolution of 1.5 dB has been assessed for the tag with coupled microstrip dipoles which allows for a significant increase in the coding capacity.

Tags with ground planes appear to be better candidates for implementing RCS synthesis due to their greater simplicity. The presence of the ground plane makes it possible to isolate the tag from the object on which it is positioned. They also have higher intensity responses. Different methods to control the magnitude have been studied and it appears clearly that the simplest method consists of modifying the orientation of the resonators individually with respect to the antennas. This method will be used in the next section for the RCS synthesis.

RCS Synthesis

5.1. Introduction

The identification technologies market is currently dominated by optical barcodes and traditional RFID. Although it combines the advantages of both of these technologies, chipless RFID has struggled to become a serious alternative. One of the major roadblocks is the limited quantity of information encoded by chipless RFID tags at this time. The current memory capacity does not exceed about 50 bits, but a minimum of 128 bits is necessary to hope to penetrate the mass market. The tag must also retain the dimensions of a credit card (8.5 cm ×5.4 cm). This objective corresponds to a surface coding density of 2.8 bits/cm².

This chapter presents a method of designing tags whose coding is based on the overall form of the tag. This approach is promising in terms of coding capacity and makes it possible to get closer to the objective. Two different cases will be studied. The first case is composed of resonant structures with a ground plane and the second case with low resonant structures without a ground plane. We will see that for the broadband case, the main difficulty is related to couplings that appear between the different structures.

5.1.1. Coding on the appearance of the response

The most common coding technique in chipless RFID is frequency positioning of peaks present in the signature of the tag (see section 2.3 in Chapter 2). Each peak physically corresponds to a resonant structure in the tag, and therefore there is a direct relation between the quantity of information encoded and the number of resonators present in the tag.

However, signals have more information than the presence of a peak at a given frequency. For instance, Figure 5.1 shows the different distinctive elements of a signal that can be exploited to code information. Specifically, we can observe the position, magnitude or width of a peak, position of a dip, or slope of the curve at a given point. When they are present in the signature of a tag, these characteristic elements are the result of physical phenomena or characteristics related to elements of the tag. The objective is therefore to determine a geometric parameter that makes it possible to control one of these distinctive elements in order to code information on it. The success of frequency position coding is explained by the fact that the resonant frequency proper to a resonator is an intrinsic parameter of the tag that does not depend on either the read range or the form of the excitation [REZ 15a]. The quality factor (or simply Q factor) associated with the resonance is also an intrinsic parameter. However, it is more difficult to control than the resonant frequency. For the time being, there are no examples of chipless tags whose information is encoded using the Q factor.

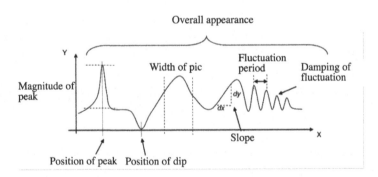

Figure 5.1. *Particularities of a signal that can be used to code information*

However, any signal contains much more information than what can be deduced from only the presence of a peak at a given frequency. Rather than focusing on certain specific elements of the signature for coding, a more general approach consists of exploiting the overall appearance of the signature (indicated in Figure 5.1) to code information. In a certain way, this approach includes the previous approaches because it takes into account all of the particularities of the signal. The difficulty then rests in the exact synthesis of the entirety of the response.

One way to assess the quantity of information contained in any signature is to convert the analog signal into a digital signal. The difference compared to a classic analog-to-digital converter is that here we are working in the frequency. As illustrated in Figure 5.2, the signal is sampled according to a step δF, which transforms the continuous-time signal into a discrete-time signal. The magnitude of the signal is then quantified for each sample according to a magnitude step $\delta\sigma$.

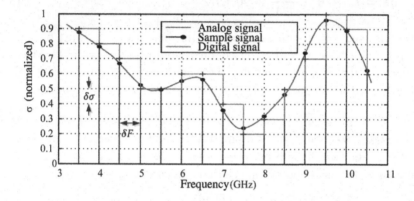

Figure 5.2. *Sampling and quantification of a signature. For a color version of this figure, see www.iste.co.uk/rance/rfid.zip*

If we consider an exploitable bandwidth ΔF, the number of samples is given by $N = \Delta F/\delta F$. Each sample is associated with a magnitude level that depends on the total magnitude variation span $\Delta\sigma$ and the quantification step $\delta\sigma$. In total, we have $M = \Delta\sigma/\delta\sigma$ different magnitude levels. If we consider that the magnitude level of a sample does not depend on the value of its neighbors, we are able to distinguish a number N_t of different signatures, with:

$$N_t = M^N = \left(\frac{\Delta\sigma}{\delta\sigma}\right)^{\Delta F/\delta F} \tag{5.1}$$

In order to evaluate the potential offered by this approach, we determine a pair of values that would make it possible to reach the objective of 128 bits. We use the UWB band (3.1 –10.6 GHz) which provides a bandwidth of

7.5 GHz. We give it $M = 3$ different magnitude levels, which makes the independence of a sample's magnitude compared to its neighbor plausible. The frequency step δF is given by:

$$ln_2(N_t) = \Delta F / \delta F \cdot ln_2(M) \qquad [5.2]$$

$$\delta F = \Delta F \cdot \frac{ln_2(M)}{ln_2(N_t)} \qquad [5.3]$$

A.N.:

$$\delta F = 7.5 \cdot 10^9 \cdot \frac{ln_2(3)}{128} = 90 \, MHz \qquad [5.4]$$

We can see that for these two realistic quantities, we must be able to detect a magnitude variation based on a frequency step of 90 MHz. In comparison, the frequency window between two consecutive peaks that is generally used for frequency position coding is in the order of 50 MHz.

For $M = 3$ different magnitude levels, a frequency step of $\delta F = 90$ MHz must be taken to code 128 bits on the UWB band. This quick study makes it possible to see that the approach of coding on the overall appearance seems promising in terms of coding capacity compared to current methods.

The approach of coding on the overall appearance of the signature radically modifies the philosophy of coding. Contrary to classic methods where a resonant structure is associated with a specific element of the signature, it appears in effect possible to code information using low or non-resonant structures. In this way, we can break away from the direct relation between the number of resonators present in the tag and the number of bits encoded. In addition, using the broadband response of low or non-resonant scatterers makes it possible to respond to realization issues in tags without ground planes printed on low-cost materials like paper. In this case, high dielectric losses ($\tan \delta \approx 0.1$) limit the selectivity of the peaks and make frequency position coding less efficient.

Frequency position coding and coding on the overall appearance are presented schematically in Figure 5.3. For classic frequency position coding, (Figure 5.3(a)), each resonator is associated with a peak in the spectral signature of the tag. For low losses (blue curve), the peaks have a high selectivity and a low spectral occupancy δF. It is therefore possible to define a large number of frequency windows for coding. For higher losses (red curve), the peaks have lower magnitude levels and are therefore less selective. This translates into a larger spectral occupancy $\delta F'$. Larger spacing between two peaks is necessary to prevent interference phenomena, which impacts the number of frequency windows usable for coding. Coding on the overall appearance (Figure 5.3(b)) exploits the response of broadband structures. It is therefore possible to use resonators with higher losses. The interference phenomena between the responses of different structures are taken into account at the time of design.

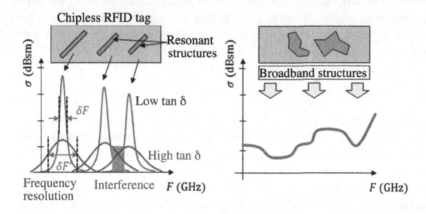

Figure 5.3. *Comparison between frequency position coding and overall appearance coding; a) effect of losses on frequency position coding; b) overall appearance coding. Low resonant broadband structures can be used. For a color version of this figure, see www.iste.co.uk/rance/rfid.zip*

Apart from the information coding aspect, the realization a tag with a specific RCS can be of interest to other domains such as reflector arrays or Fabry–Perot antennas. In these domains, it is often indispensable to associate the EM performances of a parametrized unit cell with their physical behavior like the variation of reflection-transmission coefficients in magnitude and in phase.

5.1.2. *Problem analysis*

Implementing the general form coding approach relies on the possibility of generating a chipless tag with a given EM signature. The design of such a tag comes down to solving the inverse scattering problem. This problem is generally ill-posed and too complex to be solved by a direct analytic approach [BUL 10, KAT 13]. In order to circumvent the difficulties related to solving this inverse problem, the tag will be realized using an assemblage of a certain number of unit cells whose individual frequency responses are assumed to be known. In what follows, the problem of determining the form of a tag with a specific EM signature will be designated as the "synthesis problem". This chapter focuses on developing a method of solving the synthesis problem and analyzing the results obtained. The discussions contained in this chapter can also be seen as a generalization of the problem of magnitude coding presented in the previous chapter. The problem here is more complex to the extent that the phase shift between the signals emitted from each element will play a fundamental role. We will also evaluate the possibility of using low resonant elements.

The inverse scattering problem is considered very difficult and became critically important at the time of the appearance of radar at the start of World War II. Despite significant efforts made by researchers at the time, no general solution was found. It was through the effort to circumvent the difficulties related to this problem that the Friend and Foe (IFF) identification system appeared [BRO 99], which is considered to be the forerunner of traditional RFID. Even today, after more than 50 years of research, this problem remains unsolved despite the substantial progress realized for processing inverse problems [COL 12] and the evolution of computing power. Generally, the alternative methods implemented focus on an adjusted form of the problem. For instance, in the context of radar detection [TAI 05], the form of the target is not determined from the resolution of the inverse problem but by comparing the signature measured to a database. Consequently, we will not attempt to solve the synthesis problem analytically but instead we will present an approach that makes it possible to treat the problem indirectly.

The synthesis problem consists of determining a cause (the form of the tag) given an effect (the electromagnetic field diffracted by the tag), which is characteristic of an inverse problem. The synthesis problem can be considered to be a particular formulation of the "inverse scattering problem" but for which the objective field E_O is specified arbitrarily instead of resulting from the measurement of a target.

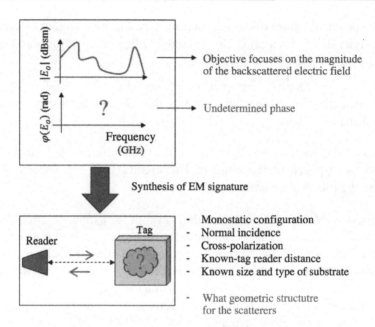

Figure 5.4. *Inverse synthesis problem. We want to determine the form of the tag in order to obtain a given diffracted field. The specifications will focus exclusively on the magnitude of the objective field. For a color version of this figure, see www.iste.co.uk/ rance/rfid.zip*

For the synthesis problem, the specifications focus on the values of the diffracted electromagnetic fields that are assumed to be known *a priori*. The synthesis problem of the tag therefore consists of determining the form of the tag responsible for these diffracted field values (see Figure 5.4). The "objective" signature can be specified for a given measurement configuration, which makes it possible to consider only the unknowns related to the target itself.

In the context of chipless RFID, the uses generally adopted allow us to simplify the problem:

– we consider a flat target whose dimensions are generally comparable to the wavelength;

– we consider the case of a monostatic measurement, with a target under normal incidence, which corresponds to most applications;

– the physical nature of the tag (permittivity and loss tangent of substrate, conductivity of conductors) is also assumed to be known at the start and will therefore be included in the input data;

– the read ranges are generally set by the application. The electromagnetic simulations presented in this chapter correspond to a tag-reader distance of 1 m.

If we come back to the theoretical aspect, we observe that the synthesis problem is ill-posed in the sense of Hadamard [HAD 14]. Indeed, we can mention the following elements:

Existence of the solution: In the case of a synthesis problem, the diffracted field E_O does not result from the measurement of a physical target but rather constitutes an "abstract" objective. In these conditions, nothing guarantees that the objective field E_O is actually realizable, in particular if the magnitude and the phase of the diffracted field are chosen without considering the physical laws of diffraction. The ability to determine a coherent object therefore demands a good prior knowledge of the direct problem. In order to set realistic magnitude levels, we will use measurements of existing tags as a base.

Unicity of the solution: Nothing guarantees the unicity of the solution: for a given electromagnetic field value, different targets may have caused this field. Contrary to other inverse problems unicity is not a critical problem for synthesis.

In addition, generally, we will consider that the objective focuses exclusively on the magnitude and not on the phase which will remain arbitrary. There are several reasons for this hypothesis that are related to certain specificities of the direct problem. First of all, it is difficult to obtain a physically realizable objective if the magnitude and the phase are imposed independently of one another. Indeed, physics imposes a connection between the phase and the magnitude of the diffracted field, and this link is not easy to grasp; it generally demands a good knowledge of the direct problem. In addition, when we measure EM fields in practice, the slope of the phase depends directly on the read range (see Appendix A). This final point is problematic because the synthesis of the objective then depends on a parameter outside the tag, which we are trying to avoid. When the phase of an objective is not specified, we are talking about an arbitrary phase problem [BUL 10].

5.1.3. *Principle of the resolution method*

A general resolution approach to the synthesis problem consists of associating a shape generation algorithm with electromagnetic simulation software used in conjunction with a genetic algorithm optimization. This technique has been used successfully to design non-imposed UHF RFID tags for broadband and robust applications for an environment that is unknown at the time of tag design [CHA 11]. This approach is somewhat different from a classic antenna optimization to the extent that no tag topology is set a priori, only the manner of generating a shape is coded. This type of approach appears possible given the current computation capacities of computers but the computation time is still prohibitive. In addition, this type of approach only provides a limited understanding on the physical systems involved. It is not possible to easily deduce the design rules that would make it possible to start from a solution that is closer to the expected one. This approach has therefore been excluded in favor of semi-analytic approaches.

The method chosen for the synthesis comes down to decomposing the signature of the objective on a basis of structures whose signatures are known individually. It corresponds in a certain way to the orthogonal projection of the objective on the space formed by the response of different structures that will be chosen as elements of the basis. This linear approach does not take into account the electromagnetic couplings that appear between the elements. The couplings are taken into account in a second stage. The principle of the resolution method is represented schematically in Figure 5.5. It is composed of five main steps that are indicated by the letters A to E in the diagram. The steps are presented briefly here and will be reviewed in detail in the rest of the chapter.

Basis of resonators

The first step consists of choosing the basis of resonators. The EM signature (backscattered field) of each isolated element $R_i(\omega)$ is obtained by simulation (resolution of the direct problem). One benefit of this approach is the ability to reuse the structures that have already been studied for the choice of a basis family. For instance, we will begin our study by choosing the coupled microstrip dipoles that were studied in the previous chapter as a basis family.

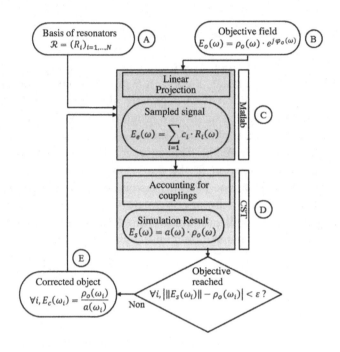

Figure 5.5. *Design algorithm of a chipless RFID tag whose coding is based on the overall appearance of the signature*

Determining the objective

The objective relates to the magnitude of the field backscattered by the target $|E_o(\omega)|$. To obtain a realizable objective, it is useful to set it after having determined a decomposition basis. In this way, we can give ourselves reasonable objective limits (magnitude).

Projecting the objective on the basis

Here we do not take into account the effect of couplings generated between the elements of the basis. This way, we have a linear problem that can be resolved quickly using a calculation software like Matlab. The operations realized on the basis elements must be able to be translated into geometric manipulations of the tag. The operations used when realizing the linear combination of two signals are the multiplication by a scalar and addition. The principles considered for the approach are described below.

Field additivity: We can consider the structures present in a tag to be isolated point sources when they are subjected to an EM field. According to this hypothesis, there is no coupling between them and the principle of superimposed fields applies, such that there is an additivity of the electric fields reflected by each structure.

Multiplication by a scalar (complex):

$$c_i \cdot R_i(\omega) = \rho_i \cdot e^{j\varphi_i} \cdot R_i(\omega) \tag{5.5}$$

This can be decomposed into two distinct physical problems:

1) Multiplication by a positive real number:

$$\rho_i \cdot R_i(\omega) \tag{5.6a}$$

Multiplication by a positive real number consists of physically modifying the magnitude of the field reflected by the structure. Different techniques were presented in Chapter 4 to realize this operation.

2) Phase shift:

$$e^{j\varphi_1} \cdot R_i(\omega) \tag{5.6b}$$

This operation is difficult to realize in practice. We could consider distancing the elements from one another, but this approach is difficult to carry out in the case of the identification issue. Indeed, this would lead to the design of 3D-tags. Nevertheless, it may make sense for addressing other issues such as minimizing the RCS of a target. We note that concretely, varying the distance does not translate the phase but modifies the slope (section 5.5.1 Appendix A).

An orthogonal projection can be defined using a norm that makes it possible to quantify the relevance of the solution, or in other words, the difference between the solution found and the objective. On a family of resonators, the projection is unique. It is also the best approximation of the signature on the family.

Accounting for couplings

After having determined the decomposition and the corresponding geometric parameters, the simulation of the tag with full-wave software (CST Microwave Studio) makes it possible to take couplings into account. This is the non-linear aspect of the problem.

Error compensation

The error between the objective and the electromagnetic simulation result is evaluated and then subtracted from the initial objective. This way, we obtain a "corrected" objective that takes into account the effects of couplings. This corrected objective is then decomposed on the basis elements in turn. This step can be repeated several times to refine the result but we will only carry out a single iteration to limit the computation time. During the first iteration, we move from the assumption that there are no couplings between the resonators to a less strong assumption, according to which the couplings do not depend on the orientation of the resonators. To adjust the magnitude of the response of the resonators, we will use an approach introduced in Chapter 4, namely controlling the orientation of the structure in relation to the tag. At each step the orientation of the resonators (and therefore the couplings) varies but in a less pronounced way, which should facilitate the convergence of the approach.

5.2. Sampling method

First, we propose decomposing the objective signal on a family of resonators with a high selectivity. This is a simplified study framework compared to "broadband" cases for two reasons. First of all, this makes it possible to reuse known resonators such as the coupled microstrip dipoles studied in the last chapter. In addition, as a first approximation, a resonant signal has a zero response outside its resonance area. Consequently, the elements of the family of decomposition can be considered to be orthogonal (with finite and non-contiguous support) which facilitates the decomposition.

On the other hand, the decomposition on a family of selective resonators presents application limits compared to "broadband" cases. The signals that can be generated by a family of resonators are made up of peaks and the synthesized signature will consequently be a sampled version of the objective (Figure 5.2). This example does not therefore provide a notable improvement to the coding capacity compared to a magnitude coding like the one presented in the previous chapter. However, the philosophy is very different. In this case, we are trying to get as close as possible to an objective. The central interest is the design protocol that could be generalized in the case of a broadband family. Meanwhile, magnitude coding concentrates on the study of resonators, which is secondary here. The "sampling method" can be seen as a first step for the broadband case aimed at concretely determining the difficulties related to design but in a simplified framework.

5.2.1. *Preliminary version of the design algorithm*

A first version of the design algorithm for the sampling method is implemented. The different error sources will be examined and will be compensated in a second stage.

5.2.1.1. *Study of a single resonator*

Here, we consider the direct problem applied to resonators. The coupled dipoles (Figure 5.6) that were presented in the last chapter are used as basis elements for the decomposition. They have a ground plane and operate in cross-polarization. That is why we only consider the cross-polarization response of dipoles, which is much easier to detect in practice. The coupled dipoles are composed of five dipoles of length L and width W separated by a gap g and oriented at angle θ (Figure 5.6(a)). We saw in the last chapter that the length L makes it possible to adjust the resonant frequency of the dipoles and that the orientation θ makes it possible to adjust the magnitude of the backscattered field. To implement the sampling method, the parameters g and W are set at $g = 0.5$ mm and $W = 2$ mm. L and θ are therefore the only variable parameters. The presence of a ground plane guarantees a good isolation from the support. The backscattered field is polarized according to the dipole axis and the dipoles are initially oriented at angle $\theta = 45°$ in relation to the vertical to maximize the component in cross-polarization.

The response of the coupled dipoles (cross-polarization component of the backscattered field) is presented graphically in Figure 5.6(b). In the vicinity of the resonance, the coupled dipoles behave like a second-order resonator with the following transfer function:

$$S_{vh}(\omega) = \frac{A}{1+jQ\left(\frac{\omega}{\omega_0}-\frac{\omega_0}{\omega}\right)}$$

[5.7]

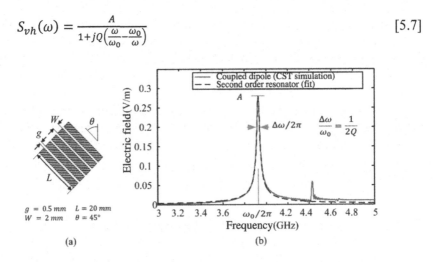

Figure 5.6. *Resonators used as basis structures for the sampling method; a) geometric parameters; b) frequency response (cross-polarization) and parametric identification. For a color version of this figure, see www.iste.co.uk/rance/rfid.zip*

A comparison between the model and the simulation is also presented in Figure 5.6(b). We observe a good agreement in the vicinity of the resonance. A parasitic resonance appears at 4.4 GHz that is not accounted for by the model and which is related to the low currents that propagate along the width of the dipoles. The model also does not account for the residual field observed after the resonance in simulation. For example, at 4.8 GHz, the magnitude given by the model is 0.004 V/m while we find 0.011 V/m in simulation.

The identification between the simulated response and the model [5.7] makes it possible to quickly establish diagrams relating the geometric parameters of the resonator to its resonance characteristics. The resonator based on gap-coupled dipoles is identified by its geometric parameters L, g, W and the component S_{vh} is obtained through electromagnetic simulation. The response obtained in simulation is fitted at the least square sense by the transfer function [5.7] to identify the characteristic parameters of the resonance: ω_0, A and Q. An example of parametric identification is illustrated in Figure 5.6. When $L = 20$ mm, we obtain the values $F_0 = 3.92$ GHz, $A = 0.28$ V/m and $Q = 163$. The diagrams represented in

Figure 5.7 have been established using nine electromagnetic simulations where L varies. These diagrams are then used to dimension the resonators.

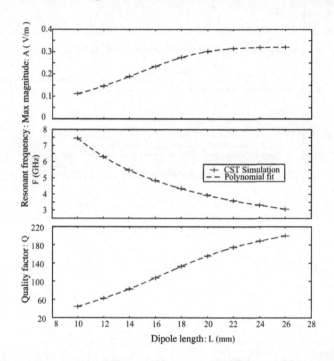

Figure 5.7. *Evolution of parameters A, F and Q with respect to the length L of the coupled dipole resonator (Figure 5.6). The simulations (CST Microwave 2014, time solver) have been performed for a substrate of dimensions 5.4 cm×5.4 cm, thickness 0.8 mm and permittivity $\varepsilon_r = 3.55$. The backscattered field is calculated for a tag-reader distance of 1 m. For a color version of this figure, see www.iste.co.uk/ rance/rfid.zip*

5.2.1.2. *Study of the basis*

The transfer function [5.7] is used to model the response of the resonators. We consider a projection basis $\mathcal{R}$ composed of $N = 8$ resonators, which corresponds to the maximum number of coupled dipoles that it is possible to integrate in a tag with the dimensions of a credit card [9]. The resonators are evenly distributed on the frequency range $I = [4.2 \text{ GHz}, \quad 6.2 \text{ GHz}]$. The response of the basis elements (isolated, no couplings) is represented in Figure 5.8. The resonant frequency and the associated length of each dipole is given in Table 5.1.

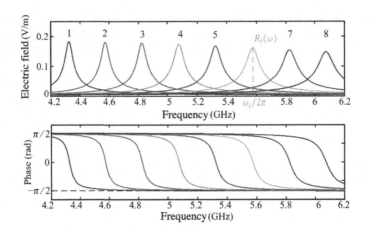

Figure 5.8. *Projection basis for the sampling method. The responses are given for isolated resonators. For a color version of this figure, see www.iste.co.uk/rance/rfid.zip*

Number of resonators	1	2	3	4	5	6	7	8
Resonant frequency (GHz)	4.32	4.57	4.82	5.07	5.32	5.57	5.82	6.07
Length L (mm)	18.1	17.0	16.1	15.2	14.4	13.7	13.1	12.5

Table 5.1. *Resonant frequency and length of dipoles of the projection basis*

The frequency response of the resonator with natural angular frequency ω_i is noted $R_i(\omega)$. The maximum magnitude A_i (given by the diagram in Figure 5.7) is obtained at the resonance ω_i:

$$\forall i \leq N,\ max_{\omega/2\pi \, \epsilon \, I} \, |R_i(\omega)| = |R_i(\omega_i)| = A_i \qquad [5.8]$$

It is instructive to evaluate the general term $|R_i(\omega_k)|$ in order to quantify the impact of a resonator on the response of its neighbors. This comes back to studying the matrix G:

$$G = \begin{bmatrix} |R_1(\omega_1)| & \cdots & |R_1(\omega_N)| \\ \vdots & & \vdots \\ |R_N(\omega_1)| & \cdots & |R_N(\omega_N)| \end{bmatrix} \qquad [5.9]$$

The maximum of the non-diagonal elements of the matrix G is found when $i = N$ and $k = N - 1$ and gives $|R_N(\omega_{N-1})| = 0.029$ V/m. The minimum of the diagonal elements of G is found when $i = k = N$ with $A_N = 0.147$ V/m. By examining the response in phase (Figure. 5.8), we can also see that the resonance takes place when the phase is cancelled out. For resonators with a high quality factor, the phase varies rapidly around the resonance to reach a value of $\pm \pi/2$. At the time of resonance, the contribution of the neighboring resonators is therefore in phase quadrature. The magnitude shift related to the contribution of a resonator on its neighbor is therefore in the worst case:

$$\| A_N + j \cdot |R_N(\omega_{N-1})| \| = \sqrt{A_N^2 + |R_N(\omega_{N-1})|^2} \approx A_N + \frac{|R_N(\omega_{N-1})|^2}{2 \cdot A_N} \qquad [5.10]$$

And consequently, there is a relative shift in the maximum magnitude of:

$$\frac{|R_N(\omega_{N-1})|^2}{2 \cdot A_N^2} \simeq 2\% \qquad [5.11]$$

This value appears sufficiently low to assume as a first approximation that the resonators are independent, that is:

$$|R_k(\omega_i)| \simeq 0, \text{ when } k \neq i \qquad [5.12]$$

5.2.1.3. Objective

The objective only focuses on the frequency variation in the module of E_0. We determine a mask whose role is to guarantee that the objective is realizable. The upper limit corresponds to the maximum magnitude of isolated resonators (in blue in Figure 5.9). It is found using the diagram given in Figure 5.7. We also consider a lower limit that is set at 0.05 V/m (in purple in Figure 5.9) in order not to integrate resonators with a response that is too low to be measured accurately.

The "objective" curve is generated by the interpolation of four points chosen randomly inside the mask (red curve in Figure 5.9). The sampling method is tested on four different objectives numbered 1 to 4 represented in Figure 5.10.

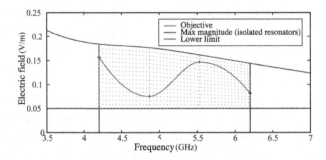

Figure 5.9. *Visualization of objective 1. The upper limit corresponds to the maximum magnitude obtained by isolated resonators. The lower limit is set at 0.05 V/m. For a color version of this figure, see www.iste.co.uk/rance/rfid.zip*

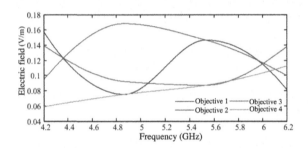

Figure 5.10. *Visualization of objectives 1 to 4 used to evaluate the sampling method. For a color version of this figure, see www.iste.co.uk/rance/rfid.zip*

Given the selective response of the elements chosen for the decomposition basis, it is clear that the signal synthesized using this approach will be a sampled version of the objective. The continuous objective can be discretized in the form of a vector of dimension N:

$$\Omega = (|E_o(\omega_1)|, |E_o(\omega_2)|, \dots , |E_o(\omega_N)|) \qquad [5.13]$$

Where the angular frequency of the sampling $(\omega_i)_{i \leq N}$ corresponds to the angular frequency of the basis elements.

5.2.1.4. Projection on the basis

This step is the linearized part of the problem (indicated by the letter C in Figure 5.5) and can be treated either analytically or with the aid of a digital calculation software like Matlab. We are trying to determine a family of positive real coefficients $(c_k)_{k \le N} \in \mathbb{R}_+{}^N$ that allow for the projection of the "objective" curve on the basis composed by the N resonators. As noted previously, it is preferable that the coefficients be real because multiplication by a positive real number translates simply in geometric terms by the rotation of the resonator. Here, obtaining real coefficients is facilitated by the hypothesis of independence [5.12].

Consider E_e the linear combination of resonators defined by:

$$E_e(\omega) = \sum_{k \le N} c_k \cdot R_k(\omega) \qquad [5.14]$$

We assume that E_e is a sampled version of the objective if we have an equality between E_e and the objective E_O for the sampling frequencies ω_i, that is:

$$\forall i \le N, |E_e(\omega_i)| = |E_o(\omega_i)| \qquad [5.15]$$

Yet, using [5.14] and exploiting the simplifying hypothesis [5.12] we have the relation:

$$|E_e(\omega_i)| = |\sum_{k \le N} c_k \cdot R_k(\omega_i)| = |c_i| \cdot A_i \qquad [5.16]$$

Consequently, a sampled version of the objective is obtained for the family of positive real coefficients:

$$c_i = \frac{|E_o(\omega_i)|}{A_i} \qquad [5.17]$$

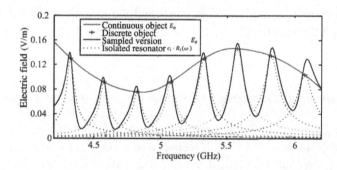

Figure 5.11. *Decomposition of objective 1 on the basis of resonators. For a color version of this figure, see www.iste.co.uk/rance/rfid.zip*

The result of the decomposition realized with Matlab using [5.17] is represented in Figure 5.11. Firstly, we note that there is indeed co-location between the objective and the isolated resonators $c_k \cdot R_k(\omega)$ (dashed line in Figure 5.11). However, for the overall response we observe magnitude values that slightly exceed the "objective" curve. A maximum exceedance of 0.025 V/m (25%) is obtained for the resonator $N = 8$. We also observe a frequency deviation between the peaks of the isolated resonators and the peaks of the overall response. The largest deviation (17 MHz) is obtained for the resonator $N = 8$. These exceedance and deviation phenomena are directly related to the approximation realized [5.12] according to which there is no influence from one resonator to another. In practice, the contribution of a resonator at the frequency of its neighbor is not zero, which modifies the expected result. This phenomenon will be accounted for in a more developed version of the sampling method. The simplification used makes it possible to find an extremely simple analytical relation expressing the magnitude value to impose on each resonator and the "objective" function. To the extent that the couplings will modify the general behaviors of the signature obtained, this first step has turned out to be very relevant.

5.2.1.5. Accounting for couplings

This step corresponds to the non-linear part of the problem. It is addressed with a full-wave simulation software (CST). To realize the simulation of the physical system, the coefficients c_i established previously must be translated in terms of geometric parameters. We showed in the last chapter that the multiplication of a resonator's response by a positive real number can be translated geometrically by a rotation of the resonator axis with respect to the antennas. For coupled dipoles, we showed that the relation between the angle of rotation θ and the magnitude of the response comply with the equation:

$$|S_{hv}(\theta, \omega)| = |\cos \theta \cdot \sin \theta \cdot S_{vv(\theta=0)}(\omega)| \qquad [5.18]$$

Where the coefficient $\cos \theta \cdot \sin \theta$ does not depend on the frequency.

The maximum magnitude level of the peak obtained in cross-polarization is obtained at angle $\theta = 45°$ with respect to the vertical for antennas oriented along the horizontal or vertical axes. These values correspond to the ones given by the characteristics (Figure. 5.7):

$$\max_\theta |R_i| = A_i = |S_{hv}(45, \omega_i)| = S_{vv(\theta=0)} /2 \qquad [5.19]$$

By expressing $S_{vv(\theta=0)}$ as a function of A_i and by linearizing the trigonometric terms in [5.18] we get:

$$|S_{hv}(\theta, \omega)| = A_i \cdot \sin 2\theta \qquad [5.20]$$

By multiplying [5.19] by the coefficient c_i and using [5.20] we have:

$$c_i \cdot R_i(\omega_i) = c_i \cdot A_i = |S_{hv}(\theta_i, \omega_i)| = A_i \cdot \sin 2\theta_i \qquad [5.21]$$

From which it is possible to extract the angle θ_i that corresponds to the coefficient c_i:

$$\theta_i = \frac{\mathrm{asin}(c_i)}{2} \qquad [5.22]$$

Equation [5.22] is used to adjust the magnitude of the resonators separately by adjusting their orientation. In the context of the simplification used, it is possible to directly express the link between the angle and the value of the objective function:

$$\theta_i = \frac{\mathrm{asin}(|E_o(\omega_i)|/A_i)}{2} \qquad [5.23]$$

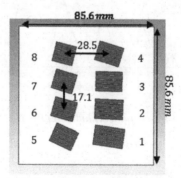

Figure 5.12. *Simulated chipless tag corresponding to objective 1. For a color version of this figure, see www.iste.co.uk/rance/rfid.zip*

The simulated tag corresponding to objective 1 is represented in Figure 5.12. A substrate with the dimensions 85.6 mm × 85.6 mm (larger dimensions than a credit card) is used to limit the edge effects. The resonators are distributed in two columns that each have four resonators. A vertical gap of 17.1 mm and a horizontal gap of 28.5 mm are used between the centers of inertia of the resonators (red dots in Figure 5.12).

The result obtained in EM simulation (CST Microwave Studio) is noted E_s and is compared to the objective backscattered field in Figure 5.13. E_s takes into account the effect of couplings as well as the physical position of the resonators. The results obtained through EM simulation provide higher magnitude levels than expected by the linear approach (Matlab curve). The most important differences are obtained starting at peak 5 with a maximum difference of 0.15 V/m (110 %) for peak 7. Similar behavior is observed for the other objectives considered. The magnitude levels obtained exceed the limit set in relation to the isolated resonators (dashed blue lines in Figure 5.13). Nevertheless, we recognize the overall appearance of the objective (minimum magnitude peak at 4.8 GHz and maximum around 5.6 GHz). The results obtained for the different objectives are provided in the summary table 5.2 at the end of this section.

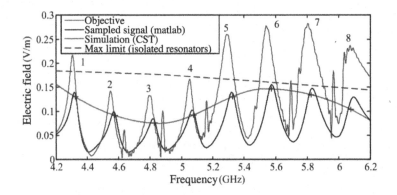

Figure 5.13. *Comparison between the result of the CST full-wave simulation and objective 1. For a color version of this figure, see www.iste.co.uk/rance/rfid.zip*

In order to compare the performances obtained for different objectives, we are introducing, in addition to the maximum difference, a measurement corresponding to a mean relative difference between the objective and the simulation:

$$\Delta_{mean}^2 = \frac{\sum_i(|E_s(\omega_i)| - |E_o(\omega_i)|)^2}{\sum_i|E_o(\omega_i)|^2} \qquad [5.24]$$

For objective 1, we find a mean difference $\Delta_{mean} = 93\%$, which represents a significant error. Accounting for the couplings will make it possible to reduce this error substantially.

5.2.1.6. Error compensation

The error present between the objective and the simulation result is primarily due to existing couplings between the resonators. We will now attempt to compensate for the effect of couplings by adjusting the orientation of the resonators. Currently, we are combining errors related to couplings with errors related to approximations realized during projection.

Let us begin by defining the error function like this:

$$E_o(\omega) - E_s(\omega) = e(\omega) \tag{5.25}$$

It is possible to decompose the error on the basis of resonators, such that:

$$e_e(\omega) = \sum_{k \leq N} e_k \cdot R_k(\omega) \tag{5.26}$$

We define an intermediary objective E'_o that takes into account the error related to couplings:

$$E'_o(\omega) = E_o(\omega) + e(\omega) \tag{5.27}$$

for which the sampled version is written:

$$E'_e(\omega) = \sum_{k \leq N} (c_k + e_k) \cdot R_k(\omega) \tag{5.28}$$

Such that the orientation of the resonators for this new objective is given by:

$$\theta_k' = \frac{\operatorname{asin}\,(c_k + e_k)}{2} \tag{5.29}$$

It is possible to repeat the compensation operation starting from the simulation result E_s' obtained for this new configuration. From a physical standpoint, the first iteration makes it possible to transition from the hypothesis according to which there are no couplings between the resonators, to a less strong assumption which stipulates that the couplings do not depend on the orientation of the resonators. As the error term decreases from one iteration to the next, the angular corrections applied to the resonators will be lower and lower and the couplings even less impacted. A local minimum is obtained when the error term stops diminishing and this constitute a stop criterion for the algorithm. However, each iteration requires a full-wave simulation with time-consuming computation. Indeed, for the multiple resonances distributed on a large frequency band, it is necessary to use a time solver. The disadvantage of working with strongly resonant structures is that all of the remaining energy in the system at the end of the simulation is related to the resonance and has a significant impact on the magnitude of the response obtained. Digital methods such as the implementation of AR filters make it possible to account for the energy that is still stored by the resonant system at the end of the calculation window but these methods only produce satisfactory results in the case of a single resonance. Very long simulations are therefore necessary to obtain good accuracy.

Consequently, we limit this study to the first iteration, which provides the most significant improvement. For the iteration n, the error $e^{(n)}(\omega)$ must be calculated in relation to the intermediary objective of the iteration $n - 1$ and not in relation to the initial objective. The intermediary objective $E_o^{(n)}(\omega)$ is given by:

$$E_o^{(n)}(\omega) = E_o(\omega) + \sum_{k=0}^{n} e^{(n)}(\omega) \qquad [5.30]$$

This way, the intermediary objective $E_o^{(n)}$ remains constant from one iteration to the next when the error $e^{(n)}(\omega)$ tends toward 0. This condition is necessary for the convergence of the algorithm.

It is important to note that in certain cases, the addition of an error can lead to unrealistic intermediary objectives, for example if the magnitude exceeds the maximum magnitude of a resonator.

The simulation results for the first iteration corresponding to objective 1 are presented in Figure 5.14.

The results obtained after the first iteration are closer to the objective compared to the first simulation. We observe a maximum deviation of 0.03 V/m (28%) for peak 8 and a mean deviation of $\Delta_{mean} = 16\%$. This result is satisfactory to the extent that the linear approach (Matlab) gives a result with an error of 11%. The results obtained for the other objectives are summarized in Table 5.2. The performance of the algorithm at each step is comparable for the different objectives.

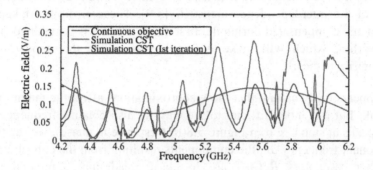

Figure 5.14. *Comparison of the simulation results obtained during the first CST simulation and after the first iteration of the compensation loop. For a color version of this figure, see www.iste.co.uk/rance/rfid.zip*

		Objective 1	*Objective 2*	*Objective 3*	*Objective 4*
Matlab	Δmax (V/m)	0.026	0.029	0.015	0.015
	Δmax relative	25%	26%	12%	14%
	Δmean	11%	11%	8%	10%
CST Simulation	Δmax (V/m)	0.15	0.17	0.14	0.11
	Δmax relative	110%	130%	113%	115%
	Δmean	94%	95%	92%	94%
CST compensation (iteration 1)	Δmax (V/m)	0.029	0.040	0.022	0.028
	Δmax relative	28%	35%	18%	26%
	Δmean	16%	14%	14%	18%

Table 5.2. *Summary of the sampling method for different objectives*

5.2.1.7. *Error sources*

The approach used above employs several simplifying approximations that have made it possible to use an analytical approach for the decomposition. Firstly, the model adopted for the resonators is a pure bandpass which does not entirely agree with the simulation results of the resonator, because we can see in Figure 5.6(b) that there is a certain residual magnitude level after the resonance. In addition, we estimated with [5.12] that the resonators would not provide a contribution to the resonant frequency of their neighbor, which is also false in light of the coefficient values of the interdependence matrix G [5.9]. These errors are taken into account and compensated during the first iteration but it is possible to correct them earlier, which will make it possible to isolate the impact of the couplings more clearly.

It appears fairly natural that these approximations affect the magnitude of the peaks but it is more difficult to see why we also obtain a frequency shift in the peaks at step C. From Figure 5.8(b) (response in phase), we see that at the resonant frequency ω_i, the components resulting from the contribution of the other resonators $R_k(\omega_i)$ are in phase quadrature compared to R_i (dephased by $\pm\pi/2$). The response of the isolated resonator is given by the transfer function [5.7], which can also be modeled by a parallel RLC circuit (Figure 5.15). The contribution of neighboring resonators can be modeled by a supplementary reactive element represented by the parasitic inductance L_p on the electric circuit equivalent with $Z_{L_p} = jX$. It is interesting to note that the same type of circuit can be used to model a resonant antenna charged by its feed line [JAC 09]. In the case of a mismatched antenna, it is common to observe a shift in the operating frequency.

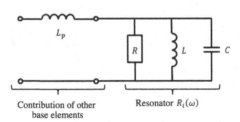

Figure 5.15. *Circuit model associated with the response of a resonator with the influence of neighboring resonators taken into account*

The transfer function associated with the response of the perturbed resonator is given by:

$$S_{vh}(\omega) = jX + \frac{A}{1+jQ\left(\frac{\omega}{\omega_0}-\frac{\omega_0}{\omega}\right)} \qquad [5.31]$$

The study of the function [5.31] effectively makes it possible to show that adding the reactive element jX causes a shift in the resonant angular frequency $\Delta\omega$ that is given by this relation:

$$\frac{\Delta\omega}{\omega_0} = \frac{1}{2Q}\,\frac{X}{R} \qquad\qquad [5.32]$$

This phenomenon is due to the fact that the summation concerns quantities that are complex (electromagnetic fields). For frequencies close to the resonance, we add two complex signals and the interference is able to move the maximum of the total signal module slightly.

5.2.2. *Improved version of the design algorithm*

A first version of the design algorithm for the sampling method has been presented and we have seen that the approximations that allowed us to realize decomposition analytically introduced error sources that can be corrected. In return, the basis can no longer be assumed to be orthogonal and consequently decomposition must be carried out using a calculation tool. The decomposition method presented here is also valid for the broadband case that will be treated next. Compared to what has been done previously, only parts B and C of the algorithm (see Figure 5.5) are modified.

5.2.2.1. Basis of resonators

We saw in the previous section that there is a residual magnitude after the resonant frequency that is not accounted for by the function used to model the resonator. The first step consists of taking into account this residual magnitude for the model of the resonator. Several different models have been tested.

The first idea uses the transfer function of a resonant high-pass filter. Unfortunately, this model does not allow for correctly approaching the simulation results. For that, the order of the filter would have to be used as a variable parameter. The resulting characteristic diagrams are therefore discontinuous and difficult to use.

The best approach consists of reusing the response of a resonant high-pass filter like in the previous section [5.7] but associating a non-resonant high-pass filter with it. The total transfer function is given by the relation:

$$S_{vh}(x) = \underbrace{\frac{A}{1+jQ(x-1/x)}}_{\substack{\text{Bandpass} \\ \text{(resonant)}}} + j\underbrace{\frac{B \cdot x^{2n}}{1+jx^n/Q'-x^{2n}}}_{\substack{\text{High pass} \\ \text{(non-resonant)}}} \qquad [5.33]$$

Where we introduce the dimensionless quantity:

$$x = \omega/\omega_0 \qquad [5.34]$$

An example of parametric identification using this model is given in Figure 5.16. We observe good agreement between the simulation result and the transfer function used.

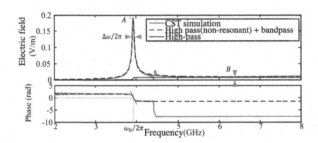

Figure 5.16. *Comparison between the response of the improved model [5.33] and the EM simulation. The parameters of the model are identified using the least squares method*

In order to limit the number of variables in the model, we arbitrarily set the value of Q' at:

$$Q' = 1/\sqrt{2} \qquad [5.35]$$

This corresponds to the limit condition in the non-resonant case and allows for the shortest possible rise time.

Still in the goal of limiting the number of variables in the model, the order of the high-pass filter is adjusted so that the slope of the phase is equal to that of the bandpass resonator around $\omega = \omega_0$, that is $\varphi = 0$ (represented by a blue segment in Figure 5.16). This way, the two signals are in phase around the resonance and there is a constructive interference. The study of the function [5.33] shows that this condition comes back to choosing the parameter n so that:

$$Q = n \cdot Q' \qquad\qquad [5.36]$$

And by accounting for the choice made [5.35], we obtain:

$$n = \sqrt{2} \cdot Q \qquad\qquad [5.37]$$

This way, the only variable high-pass parameter is B which is determined using the least square approximation between [5.33] and the simulation result.

This approach makes it possible to keep characteristics that are very similar to those obtained in the previous section. We simply have a supplementary parameter B that corresponds to the residual magnitude level that appears after the resonance. All of the variable parameters have a clear significance that can be seen directly on the response of the resonator (Figure 5.16).

The supplementary characteristic diagram of B based on the dipole length has been extracted using nine full-wave simulations and is represented in Figure 5.17. With this improved model, the other parameters have not been modified. The results presented in Figure 5.7 remain valid.

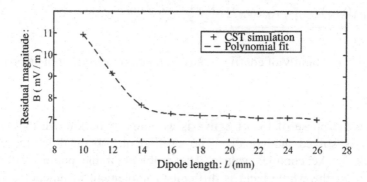

Figure 5.17. *Characteristic diagram obtained by parametric identification between the model (5.33) and 9 EM simulations. The configuration of the simulations is identical to the one described for the nomograms in Figure 5.7*

5.2.2.3. *Theoretical study*

The modeling takes into account the residual magnitude after the resonance, and contrary to the study realized in the previous section we will no longer consider that the influence of a resonator on its neighbor is zero during decomposition. Consequently, it is necessary to take into account the response of the resonators on the entire frequency band. The resolution method implemented here is therefore very similar to the method that will be used for broadband resonators, with the exception that the results are evaluated at the sampling points (discrete objective) while they will be evaluated on all points of the spectrum for broadband decomposition (continuous objective). The theoretical study is divided into three distinct problems that progress from the most general case to the most specific.

Problem 1: Complex objective, complex coefficients

Assuming that the phase of the objective is known, this comes down to finding a family of coefficients $C = (c_k)_{k \leq N}$ such as:

$$\forall i \leq N, \quad \sum_k c_k \cdot R_k(\omega_i) = E_o(\omega_i) \tag{5.38}$$

In the summation, we recognize the form of the canonical scalar product of $\mathbb{C}^N$. We can therefore write the problem [5.38] in matrix form:

$$\underbrace{\begin{bmatrix} R_1(\omega_1) & \cdots & R_N(\omega_1) \\ \vdots & & \vdots \\ R_1(\omega_N) & \cdots & R_N(\omega_N) \end{bmatrix}}_{\mathcal{R}} \cdot \underbrace{\begin{bmatrix} c_1 \\ \vdots \\ c_N \end{bmatrix}}_{C} = \underbrace{\begin{bmatrix} E_o(\omega_1) \\ \vdots \\ E_o(\omega_N) \end{bmatrix}}_{O} \tag{5.39}$$

The desired family of coefficients is obtained by inverting the matrix $\mathcal{R}$:

$$C = \mathcal{R}^{-1} \cdot O \tag{5.40}$$

When the phase of the objective is assumed to be known, this type of problem is solved analytically. The coefficients obtained by [5.40] can nonetheless take complex values. The problem of the phase shift of the resonator on the whole band is difficult to implement in practice and it is therefore preferable to find a family of real coefficients C.

Problem 2: Complex objective, real coefficients

To find real coefficients, we can rewrite the previous problem [5.39] by causing only real quantities to appear. To do this, we separate the matrices into real and imaginary parts. Re is used to denote the matrix operator that takes the real part of the matrix elements and Im is used to denote the operator that takes the imaginary part of the matrix elements. These operators do not cause the dimension of the matrix to decrease, that is:

$$\dim(\mathrm{Re}(\mathcal{R})) = \dim(\mathrm{Im}(\mathcal{R})) = \dim(R) \qquad [5.41]$$

Problem [5.39] can therefore be rewritten in the form:

$$\mathrm{Re}(\mathcal{R}) \cdot C + j \cdot \mathrm{Im}(\mathcal{R}) \cdot C = \mathrm{Re}(\mathcal{O}) + j \cdot \mathrm{Im}(\mathcal{O}) \qquad [5.42]$$

Which can be separated in the form of two independent problems:

$$\mathrm{Re}(\mathcal{R}) \cdot C_1 = \mathrm{Re}(\mathcal{O})$$

$$\mathrm{Im}(\mathcal{R}) \cdot C_2 = \mathrm{Im}(\mathcal{O}) \qquad [5.43]$$

In order to force the equality of the coefficients C_1 and C_2 we can concatenate the matrices related to real and imaginary parts, which gives the real problem:

$$\begin{bmatrix} \mathrm{Re}(\mathcal{R}) \\ \mathrm{Im}(\mathcal{R}) \end{bmatrix} \cdot C = \begin{bmatrix} \mathrm{Re}(\mathcal{O}) \\ \mathrm{Im}(\mathcal{O}) \end{bmatrix} \qquad [5.44]$$

This problem is over-determined and consequently there is no guarantee that an exact solution exists. The problem can nonetheless be solved at the least square sense in order to obtain a family of real coefficients C_{ls} such as:

$$C_{ls} = \min_{C \in \mathbb{R}^N} \left(\left\| \begin{bmatrix} \mathrm{Re}(\mathcal{R}) \\ \mathrm{Im}(\mathcal{R}) \end{bmatrix} \cdot C - \begin{bmatrix} \mathrm{Re}(\mathcal{O}) \\ \mathrm{Im}(\mathcal{O}) \end{bmatrix} \right\|^2 \right) \qquad [5.45]$$

Finding real coefficients makes the practical implementation of the solution much simpler because we already know how to modify the magnitude of the basis elements.

For this problem, the phase of the objective must be known to determine the real part and the imaginary part of the objective.

Problem 3: Real objective, real coefficients

We consider that only the magnitude of the objective is known and the problem therefore comes down to finding the family of real coefficients $C = (c_k)_{k \leq N}$ such as:

$$\forall i \leq N, \ \left| \sum_k c_k \cdot R_k(\omega_i) \right|^2 = |E_o(\omega_i)|^2 \qquad [5.46]$$

By separating the real and imaginary parts of the sum like before, we have:

$$\forall i \leq N, \ \left[\text{Re} \left(\sum_k c_k \cdot R_k(\omega_i) \right) \right]^2 + \left[\text{Im} \left(\sum_k c_k \cdot R_k(\omega_i) \right) \right]^2$$
$$= |E_o(\omega_i)|^2 \qquad [5.47]$$

Assuming that the c_k are real, we can write:

$$\forall i \leq N, \ \left[\sum_k c_k \cdot \text{Re}(R_k(\omega_i)) \right]^2 + \left[\sum_k c_k \cdot \text{Im}(R_k(\omega_i)) \right]^2$$
$$= |E_o(\omega_i)|^2 \qquad [5.48]$$

Which can be written in matrix form:

$$[\text{Re}(\mathcal{R}) \cdot C] \cdot \char`^2 + [C \cdot \text{Im}(\mathcal{R})] \cdot \char`^2 = \Omega \cdot \char`^2 \qquad [5.49]$$

Where the dot operator "$\cdot \ \char`^$" designates the element-wise power and Ω is the vector defined by [5.13] which is composed of the module of the elements of $\mathcal{O}$.

This non-linear system of equations only involves real quantities and can therefore be solved using the least square method in $\mathbb{R}$. Contrary to problems 1 and 2, it only involves the magnitude of the objective.

It is interesting to compare the results obtained for problems 2 and 3 because this gives indications about the performances that can be found for problem 2 (more general problem) when we are given an objective in magnitude and in phase.

Assuming that we solved problem 2, then we have a family of real coefficients C that minimizes [5.45]. We can assess this family as a potential solution for problem 3. According to the triangle inequality, we have:

$$\forall i \leq N, \left| \left| \sum_k c_k \cdot R_k(\omega_i) \right| - |E_o(\omega_i)| \right| \leq \left| \sum_k c_k \cdot R_k(\omega_i) - E_o(\omega_i) \right| \quad [5.50]$$

Which shows that the distance in norms between the pseudo-solution and the objective obtained for problem 3 is lower than that of problem 2. The best results are therefore expected in terms of norms when the objective only focuses on the norm (problem 3), which is understandable if we consider that problem 3 is less restricted (no constraints on the phase) than problem 2.

5.2.2.4. Results of the projection on the basis

To realize the decomposition of objective 1 on a coupled dipole basis, we use the formalism of problem 3 [5.49] that is solved using the least squares method in Matlab. We find a family of positive real coefficients (c_i) that correspond to the sampled version of the signal represented in Figure 5.18.

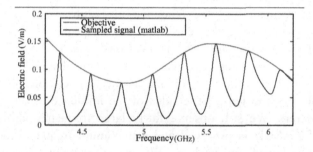

Figure 5.18. *Result of the projection on the basis by the method of least squares (Matlab) in the framework of the formulation of problem 3: real objective, real coefficients. For a color version of this figure, see www.iste.co.uk/rance/rfid.zip*

The result of the decomposition is very satisfactory. There is effectively a co-location between the "objective" curve and the maximums of the peaks in the sampled version. We observe a maximum deviation that is less than 10^{-4} V/m. This is a significant improvement compared to the analytical version because for the same objective, there was a maximum deviation of 0.025 V/m.

Having a model [5.33] that is more similar to the behavior of resonators and being able to execute a more reliable decomposition makes it possible to considerably limit the error sources related to the "linear" part of the problem (step C on the diagram of the algorithm in Figure 5.5). This way, the error sources are not accumulated when we simulate the physical configuration and account for couplings. We can therefore ensure that the shift is indeed related to the impact of the couplings and not to various approximations realized previously.

Similarly, we note that it is particularly important to carry out a valid decomposition when we choose to iterate the correction process (step F in Figure 5.5), so as not to inject additional errors with each iteration.

5.2.2.5. Accounting for couplings

Once the coefficients c_i are calculated, we determine the equivalent angles using [5.22] and the resonators are positioned on the tag in a configuration similar to the one in Figure 5.11. The response of the physical structure is obtained through EM simulation (CST). The result of the simulation is compared to the objective in Figure 5.19. We observe a maximum magnitude exceedance of 0.08 V/m for peak 5 (65%) and a mean deviation of 44%. Compared to the previous results, there is a marked improvement at this step because before, we obtained a maximum exceedance of 0.15 V/m (110%) and a mean deviation of 94%. This approach allows us to get a better idea of the influence of the presence of several scatterers. We still observe fairly large frequency shifts. A mean deviation of 27 MHz is observed and a maximum frequency deviation of 38 MHz is obtained for peak 5. We can reasonably assume that the frequency shift is due to couplings between the resonators and not modeling errors on isolated resonators.

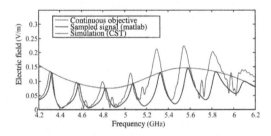

Figure 5.19. *Comparison between the results expected theoretically (Matlab, linear approach) and results obtained through electromagnetic simulation (CST, accounting for couplings). For a color version of this figure, see www.iste.co.uk/rance/rfid.zip*

5.2.2.6. *Correcting the effects of couplings*

Like in the previous case, the error on the magnitude existing between the objective and the result of the simulation is assessed and taken into account by setting an intermediary objective [5.27]. The orientation of the resonators is modified [5.28] as a consequence. The result of the simulation obtained after the first iteration is presented in Figure 5.20.

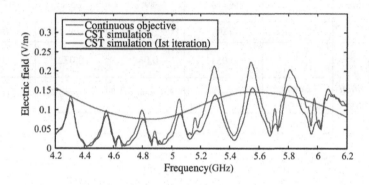

Figure 5.20. *Comparison of the simulation results obtained during the direct CST simulation and after the first iteration of the compensation loop. For a color version of this figure, see www.iste.co.uk/rance/rfid.zip*

Like in the previous case, the first iteration shows a marked improvement in the results compared to the direct simulation. The mean error Δmean passes from 44% for the direct simulation case to 15% after accounting for the couplings. A mean frequency shift of 4 MHz is observed between these two results. The results obtained during the first iteration are comparable to those obtained after the first iteration of the naïve method. For objective 1, a mean deviation of 16% has been obtained after the first step. The low improvement in terms of results compared to the naïve approach is probably due to the fact that to take into account the couplings, the magnitude of the peaks has been assessed at the level of the resonant frequency of the isolated resonators and not at the level of the maximums of the overall response. We saw in the previous section (Figure 5.19) that there could be a relatively large frequency shift between these two values. This first point explains why the first two peaks that seem to be well-positioned during the direct simulation are more distant from the objective during the first iteration. Better results could probably be obtained by evaluating the difference between the peaks and the curve at the level of the maximums of the overall response. Another explanatory element is the identification of a non-constant coupling as a function of the angle.

		Objective 1	Objective 2	Objective 3	Objective 4
CST Simulation	Δmax (V/m)	0.083	0.092	0.074	0.064
	Δmax relative	64%	59%	65%	67%
	Δmean	44%	39%	44%	48%
CST compensation (iteration 1)	Δmax (V/m)	0.051	0.062	0.032	0.021
	Δmax relative	48%	56%	25%	19%
	Δmean	15%	16%	14%	14%

Table 5.3. *Summary of results obtained using the improved sampling method for different objectives*

The results obtained for the different objectives are given in terms of maximum error and mean error in Table 5.3. The results are comparable from one objective to another.

The decomposition method presented in the framework of the improved version of the algorithm is also valid for the broadband case that will be treated in the next section. The main difference is the fact that the results will be on the entirety of the frequency band in question (continuous objective).

5.3. Decomposition on broadband structures

5.3.1. *Basis of resonators*

The elements used for the projection basis are coupled dipoles without a ground plane (Figure 5.21). The coupled dipoles without a ground plane are composed of a variable number of strips n between 1 and 4 of width $W = 2$ mm and variable length L. The strips are spaced by a gap $g = 0.5$ mm. The dipoles are positioned on a RO4003 substrate without a ground plane of thickness 0.8 mm and permittivity $\varepsilon_r = 3.55$. Like before, we are only interested in the cross-polarization response of the dipoles. The dipoles are initially oriented at angle $\theta = 45°$ to maximize the cross-polarization component of their response. The values of the backscattered fields are obtained through EM simulation (CST) using a probe placed 1 m from the tag. By identification with a second-order band-pass filter, we find a quality factor Q in the order of 2.9 for these elements. At the same resonant frequency, the dipole without a ground plane is much less selective than the coupled dipole with a ground plane for which we obtain a quality factor $Q = 80$. Given the absence of the ground plane, we no longer find the resonant cavity aspect that is responsible for the high selectivity of microstrip patches.

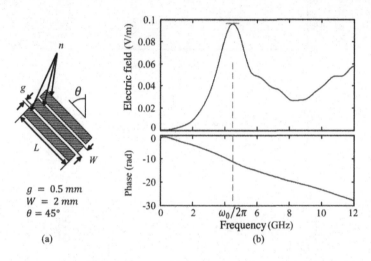

(a)   (b)

Figure 5.21. *Coupled dipoles without a ground plane used for broadband decomposition; a) geometric parameters associated with the dipole; b) response of the coupled dipoles when* $n = 3$ *and* $L = 26$ *mm. A quality factor of* $Q = 2.9$ *is obtained by identification with a band-pass filter*

A parametric study of the response of the dipoles has been realized by varying the number of strips n and the length L of the dipoles. The curves corresponding to a maximum magnitude A of the dipoles as well as to their associated resonant frequency F (Figure 5.22) have been extracted using a series of 32 electromagnetic simulations with n varying between 1 and 4 and L varying between 12 mm and 26 mm.

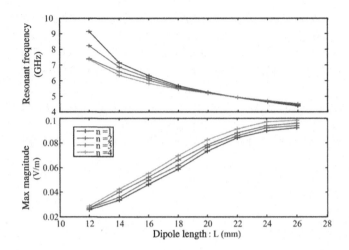

Figure 5.22. *Parametric study of the response of the dipoles without a ground plane presented in Figure 5.21. For a color version of this figure, see www.iste.co.uk/rance/rfid.zip*

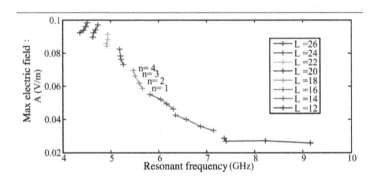

Figure 5.23. *Parameters of resonance for dipoles resulting from the parametric study. For a color version of this figure, see www.iste.co.uk/rance/rfid.zip*

These two characteristic diagrams can be combined in order to provide a better idea of the resonance parameters of the dipoles (Figure 5.23). The coupled dipoles behave as a first approximation as half-wave resonators. We can note that the magnitude of the response is generally not as high as in the case of dipoles with a ground plane.

Contrary to the case of coupled dipoles with a ground plane, it is difficult to propose a behavior model for these broadband resonators and consequently, we will use the simulation results directly to make decomposition calculations.

We consider a decomposition basis $\mathcal{L}$ composed of $N = 32$ resonators used for the parametric study. The response of each basis element will be designated by $\Lambda_i(\omega)$ in what follows to mark the difference from the highly resonant case. We are only interested in the response of the basis elements on the frequency range $I = [4.2\ \text{Ghz},\ 6.2\ \text{Ghz}]$. For the sake of readability, we will only present the basis elements for which the number of strips is $n = 4$ in Figure 5.24. The responses for other values of n are similar with a slight variation in magnitude.

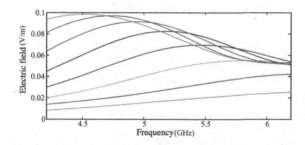

Figure 5.24. *Decomposition basis for the sampling method – broadband elements approach (Only the responses of 8 resonators with n = 4 are represented). The basis used is composed of 32 resonators in total. The responses are given for the isolated resonators. For a color version of this figure, see www.iste.co.uk/rance/rfid.zip*

The responses of the resonators vary rather slowly on the frequency range considered and, consequently, it will be difficult to synthesize responses with quick variations.

The objective will only focus on the magnitude. To facilitate the comparison, we will retain the objectives that we set in the previous part.

5.3.2. Decomposition on the basis

Mathematically, the response of the resonators belongs to the space of complex functions of a real variable. This space is naturally equipped with a Hermitian inner product:

$$\langle f|g \rangle = \int_I f(\omega) \cdot \bar{g}(\omega) \, d\omega \qquad\qquad [5.51]$$

Its associated norm will be noted by double bars to differentiate it from the complex module:

$$\|f\|^2 = \langle f|f \rangle = \int_I |f(\omega)|^2 \, d\omega \qquad\qquad [5.52]$$

Contrary to the sampling method, the response is assessed on the whole band in question and not on a discrete set of points. The objective only focuses on the magnitude of the responses. From a purely theoretical perspective, we can adopt a purely continuous formalism. We are therefore attempting to determine a family of real coefficients $C = (c_k)_{k \leq N}$ such as:

$$\left\| \left| \sum_k c_k \cdot \Lambda_k(\omega) \right| - |E_o(\omega)| \right\|^2 = 0 \qquad\qquad [5.53]$$

However, the digital signals are discrete and it is therefore possible to directly use the formalism proposed in [4.47]. We simply need to fill in this condition for all points on the frequency band and not only for the ω_i, which moreover do not have a particular sense here. Using the notation that has been introduced for the broadband structures, we have:

$$\forall \omega \in I, \ \left[\sum_k c_k \cdot \mathrm{Re}\big(\Lambda_k(\omega)\big) \right]^2 + \left[\sum_k c_k \cdot \mathrm{Im}\big(\Lambda_k(\omega)\big) \right]^2 = $$
$$|E_o(\omega)|^2 \qquad\qquad [5.54]$$

As seen previously in [5.49], this relation can be put in matrix form and be solved using the method of least squares for positive real coefficients. The decomposition is realized using Matlab by adding a constraint on the coefficients so that they can only take values belonging to the interval [0, 1]. This constraint makes it possible to avoid solutions that are difficult to realize in practice for which we have coefficients greater than one and therefore resonators that must exceed their maximum value.

Subsequently, we have chosen to take a maximum of six resonators in order to limit the dimensions of the tag. We start by carrying out the decomposition on the entire basis (32 elements) and then we only select the six elements whose coefficients are the most significant. We therefore consider the restricted basis $\mathcal{L}_r$ composed of these six elements and realize a new decomposition on this basis. This approach makes it possible to select a good decomposition basis when we have an oversized decomposition family. Once the restricted basis is determined, the iterations are realized with this same basis in order not to modify the couplings too significantly. This approach provides an optimal (linear) solution on the restricted basis but not necessarily on the complete basis.

The decomposition of objective 1 on the basis of broadband elements has been realized using Matlab. The result obtained is noted $E_l(\omega)$ and is represented in Figure 5.25. The evolution of coefficients c_k when the basis is restricted to six elements is represented in Figure 5.26.

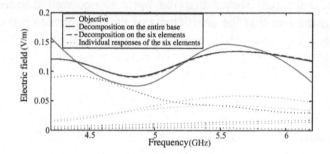

Figure 5.25. *Result of the decomposition on a basis of broadband resonators – objective 1. For a color version of this figure, see www.iste.co.uk/rance/rfid.zip*

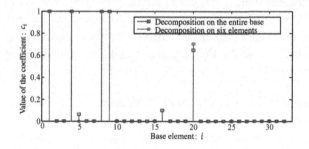

Figure 5.26. *Evolution of the coefficients for a basis restricted to six elements – objective 1. For a color version of this figure, see www.iste.co.uk/rance/rfid.zip*

We can see in Figure 5.26 that the starting basis is relatively poorly adapted to the decomposition of the objective. This translates into a projection on the basis that is rather far from the starting objective. We can define a normalized distance Δ_{mean} using the Hermitian norm between two curves which will make it possible to compare the results for different objectives. This distance is simply an adaptation to the continuous case of [5.24]:

$$\Delta_{mean} = \frac{\||E_o(\omega)|-|E_l(\omega)|\|}{\|E_o(\omega)\|} \qquad [5.55]$$

Here we obtain a distance $\Delta_{mean}= 10\%$ between the objective and its decomposition on the basis. However, better results are obtained for other objectives (1%, 3.5% and 3% for the objectives 1, 2 and 3 respectively). Another indicator that shows that the basis is not well adapted to the projection is the fact that the coefficients of the decomposition take values corresponding to the maximum of their variation range. It should be noted, however, that the restriction of the basis only very slightly degrades the result of the decomposition, in comparison with the decomposition on 32 elements.

5.3.3. *Accounting for couplings and corrections*

Like in the case of the coupled dipoles with a ground plane, for a particular orientation of the dipoles $\theta = 0$, the polarimetric scattering matrix S is zero with the exception of the component S_{vv}. The other components of the matrix S are less than -120 dB in simulation. Consequently, the rotation of the dipoles without a ground plane influences the magnitude of the backscattered field according to the simplified equation [5.18]. The coefficients obtained from the decomposition are translated in terms of angles θ_i using [5.22].

Let us consider a tag with large dimensions (Format A4: 210 mm × 297 mm) in order to space the resonators to limit the effect of couplings which is very significant for broadband resonators. A spacing of 74 mm between two neighboring resonators is sometimes greater than $\lambda_r/2$ which can translate into destructive interferences as in the case of an antenna array. This phenomenon is particularly problematic for broadband (λ_r variable) because the phase shift between the different responses varies based on the frequency. If we set the position of the resonators first and consider them to be point sources, it is possible to account for this effect during the decomposition which will allow for more compact tags than the method proposed below. The resonators are positioned in a circle around a central point (aligned with the field measurement sensor) so that the response of different resonators adds up in phase on the whole frequency band. The structure of the simulated tag is represented in Figure 5.27.

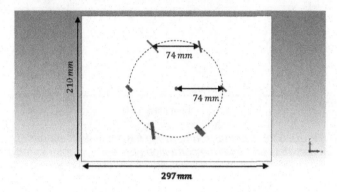

Figure 5.27. *The broadband resonators are positioned on a tag with the dimensions of an A4 sheet in order to consider large spacing to limit the impact of couplings. The resonators are placed in a circle so that their responses add in phase regardless of the frequency. For a color version of this figure, see www.iste.co.uk/rance/rfid.zip*

The result of the simulation of the tag is represented in Figure 5.28. Taking into account the couplings has a very significant influence on the level of the tag's response compared to the linear prediction (Matlab). A relative distance Δ_{mean} = 13% is obtained between the simulation and the objective. The couplings make it possible to obtain greater variation coefficients than for linear prediction.

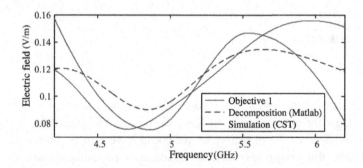

Figure 5.28. *Comparison between the response of the simulated tag (CST) and the linear response (Matlab). For a color version of this figure, see www.iste.co.uk/rance/rfid.zip*

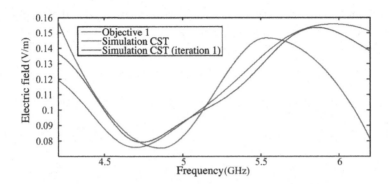

Figure 5.29. *Comparison between the simulated responses immediately and after the correction phase. For a color version of this figure, see www.iste.co.uk/rance/rfid.zip*

In the broadband case, it is difficult to view the impact of the couplings on the response of one resonator in particular because there is no dominant contribution to a given frequency contrary to the sampling approach. The method proposed in the previous sections [5.27] remains useful nevertheless. The simulation result after the first iteration is compared to the immediate simulation result in Figure 5.29. An improvement of the result for the region less than 4.7 GHz is noted but the general improvement is less marked than for the sampling method ($\Delta_{mean}=11\%$). For low resonant structures, the EM simulations are much faster than in the resonant case and it therefore appears possible to repeat the process. However, we do not observe a notable improvement of the result.

5.4. Conclusion

An algorithm for the design of chipless RFID tags whose coding is based on the overall appearance of the response has been proposed and tested in simulation for resonant and low resonant structures. The results obtained for resonant tags are encouraging. In the case of low resonant tags the results were not as close to the objective as we could have hoped. This is related to the fact that the couplings between resonators (non-linear) are markedly stronger in the non-resonant case and complementary studies of these couplings are certainly necessary to improve the results obtained. The results obtained are nonetheless sufficiently close to the objective to consider this first step an interesting starting point for the implementation of a genetic algorithm. Several degrees of freedom could be exploited to improve the response. For instance, we could consider adjusting the position of the resonators within the tag in order to modify the configuration of the couplings and thus find a tag closer to the response. The low resonant case is favorable for the use of digital methods because the associated electromagnetic simulations are relatively short (between 5 and 20 min). It is interesting to note that although this was not done in the previous example, it seems possible to semi-analytically predict the positive and destructive interferences that can exist for resonators positioned randomly, in a similar way to what is done for antenna arrays. This point constitutes a second degree of freedom that is potentially usable to improve the response of the tags.

For perspective, an important step is to test the signature synthesis method in measurement. We saw in the previous chapter that magnitude coding posed some specific problems that are also found for coding on the general form. Practical validation is an important next step to validate the approach.

5.5. Appendices

5.5.1. *Appendix A: Effect of the read range on the signature of a tag*

The radar range equation makes it possible to take into account the effect of a variation in the tag-reader distance on the magnitude of a tag's signature. Using the radar range equation, we see that:

$$|S_{11}| = \sqrt{\frac{P_r}{P_t}} = \sqrt{\sigma}\frac{G\lambda}{\left(2\sqrt{\pi}\right)^3 R^2} \qquad [5.56]$$

By considering a different read range R':

$$|S_{11}'| = \sqrt{\frac{P_{r'}}{P_{t'}}} = \sqrt{\sigma}\frac{G\lambda}{\left(2\sqrt{\pi}\right)^3 R'^2} \qquad [5.57]$$

From which:

$$|S_{11}|/|S_{11}'| = R'^2/R^2 \qquad [5.58]$$

Which shows that for the representation of the signature in decibels, a variation in the tag-reader distance simply corresponds to a translation of the objective curve (Figure 5.30(a)).

Let us now consider the phase term. Following a far-field hypothesis, the wave radiated by the transmitting antenna and the wave reflected by the tag (backscattered field) behave like spherical waves. The phase term of a spherical wave based on the propagation distance is typically given by: e^{jkR}. By expressing the wavenumber with respect to the frequency for a wave that propagates in free space, we have:

$$\varphi = 2 \cdot \pi \cdot f/c \cdot R + \varphi(f) \tag{5.59}$$

where $\varphi(f)$ is a function that represents the effects other than the propagation such as the reflection on the tag or the resonance, the phase shift induced by the antennas. This function does not depend on the read range R. The slope of the phase is directly related to the read range, which is represented schematically in Figure 5.30(b).

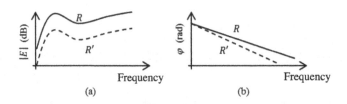

Figure 5.30. *Effect of the tag-reader distance on the response of a chipless RFID tag: a) magnitude; b) phase*

5.5.2. *Appendix B: Frequency deviation related to a parasitic reactive element*

The response of a resonator perturbed by a parasitic element is modeled by the transfer function:

$$S_{vh}(\omega) = jX + \frac{A}{1+jQ\left(\frac{\omega}{\omega_0}-\frac{\omega_0}{\omega}\right)} \tag{5.31}$$

We introduce the dimensionless quantity:

$$x = \frac{\omega}{\omega_0} \tag{5.60}$$

Equation [5.31] can be written in the form:

$$S_{vh}(x) = \frac{A+j[X(1+Q^2(x-1/x)^2)-AQ(x-1/x)]}{1+Q^2(x-1/x)^2} \qquad [5.61]$$

The resonance condition is related to the cancellation of the reactive part. We are therefore looking for values of x such as:

$$X(1 + Q^2(x - 1/x)^2) - AQ(x - 1/x) = 0 \qquad [5.62]$$

We note:

$$\alpha = Q(x - 1/x) \qquad [5.63]$$

Such that [5.61] can be rewritten in the form of a second-degree polynomial:

$$X\alpha^2 - A\alpha + X = 0 \qquad [5.64]$$

Whose roots are given by:

$$\alpha_{1,2} = \frac{A \pm \sqrt{R^2 - 4X^2}}{2X} = \frac{A\left[1 \pm \sqrt{1 - \left(\frac{2X}{A}\right)^2}\right]}{2X} \qquad [5.65]$$

When X is sufficiently low compared to A we can make a first-order approximation of the square root:

$$\alpha_{1,2} \simeq \frac{A\left[1 \pm \left(1 - \frac{1}{2}\left(\frac{2X}{A}\right)^2\right)\right]}{2X} \qquad [5.66]$$

The resonant angular frequency of the perturbed resonator ω_r can be expressed as a function of the resonant frequency of the isolated system f_0.

$$\omega_r = 2\pi f_0 + 2\pi\Delta f \qquad [5.67]$$

And if Δf is sufficiently low compared to f_0 we have:

$$\alpha_r = Q(x_r - 1/x_r) \simeq 2Q\frac{\Delta f}{f_0} \qquad [5.68]$$

By replacing α with the expression of α_r [5.68] in [5.66], we find:

$$2Q\frac{\Delta f}{f_0} = \frac{A\left[1\pm\left(1-\frac{1}{2}\left(\frac{2X}{A}\right)^2\right)\right]}{2X}$$

[5.69]

The root bearing the "+" sign will give a very high frequency shift and therefore does not have a physical sense in this context. After simplifying, we obtain:

$$\frac{\Delta f}{f_0} = \frac{1}{2Q}\frac{X}{A}$$

[5.32]

Conclusion

Although this book focused on a specific branch of RFID, it also discussed a variety of studies. This is due to the fact that chipless RFID is at the intersection between different research domains such as antennas, radar targets, resonant circuits and traditional RFID.

This book established the theoretical foundation for a new type of coding based on the overall form of the signature. This type of coding is promising from the standpoint of coding capacity and should make it possible to attain the objective of 128 bits necessary for industrial use. The use of low resonant structures makes the approach compatible with tags realized on low-cost materials like paper. This point is indispensable for the realization of low-cost chipless tags. However, additional work must be carried out to improve the ability to account for couplings between the different structures and to evaluate the feasibility of the approach in practice.

We began by comparing chipless RFID to barcodes and UHF RFID according to performance and cost. Chapter 2 presented the different coding techniques typically used in chipless RFID. This chapter showed that the highest coding capacities were obtained from tags coded in the frequency domain. Chapter 3 precisely defined what constitutes the electromagnetic signature of a chipless tag. Theoretical tools were also provided to facilitate the design and analysis of future tags.

In Chapter 4, two complete studies of chipless tags were carried out from the design to the practical measurement in a real environment. The magnitude coding method was presented and the magnitude resolution used to distinguish two consecutive magnitude levels was assessed using a series of measurements for the two configurations. We saw that for a tag without a ground plane, a minimum magnitude resolution of 3.5 dB must be used. For a tag with a ground plane, a lower magnitude resolution of 1.5 dB can be used. This study also illustrated in practice the difficulties related to magnitude coding, which depends strongly on the environment. Given the results of this study, in the case without a ground plane, it appears to be important to know the type of object on which the tags will be positioned (permittivity, thickness) in advance in order to account for it at the time of design. Techniques that allow for controlling the magnitude of the response of tags were proposed. The easiest technique to implement and which makes it possible to obtain the greatest dynamic is based on the orientation of the tag with respect to the polarization of the antennas.

The method of controlling the magnitude of the response was reused in Chapter 5 to design tags whose coding is based on the overall appearance of the signature. Two studies were conducted. The first study included resonant structures with a ground plane, similar to the tags used in Chapter 4. Although it did not offer a coding increase compared to the case in Chapter 4, this first study enabled the use of a simplified framework (orthogonal base) for the implementation of the design method. Encouraging results were obtained in simulation for the strongly resonant case. The design method was tested for a decomposition base formed of low resonant elements. Relatively satisfactory results were obtained although accounting for couplings remains problematic.

Perspectives

The short duration of the electromagnetic simulations for the low resonant case makes it possible to consider using the results obtained in Chapter 5 as a starting point for a genetic algorithm that plays on the degrees of freedom that have not yet been exploited. For instance, it seems relatively simple to modify the positioning of the resonators within the tag in order to adjust the relative phase of the elements. By modifying the position of the elements, we also change the configuration of the couplings. In the absence of an analytical model, this approach must be conducted digitally.

Apart from the information coding aspect, the generation of a tag with a specific RCS could also be of interest to other domains of application such as reflector arrays and Fabry–Perot antennas.

The high sensitivity of the magnitude of a tag's response is problematic for designing chipless tags but it can be taken advantage of to realize low-cost sensors. For instance, it is possible to exploit the non-linearity of the response of a dipole in relation to the angle [4.30] to realize an angle sensor. Another application is the use of chipless tags for the electromagnetic characterization of a substrate. By positioning a tag on an object with a known thickness and by measuring the frequency shift induced on the resonance, it is possible to trace back to the permittivity of the object. By assuming that we also have a model relating the magnitude of the response to dielectric losses, it is also possible to measure the losses (tan δ) with this approach. These kinds of test sets are compatible with low-cost radar measurement devices (some of which have been developed for chipless RFID) and do not necessarily require costly laboratory equipment.

Bibliography

[APP 79] APPEL-HANSEN J., "Accurate determination of gain and radiation patterns by radar cross-section measurements", *IEEE Transactions on Antennas and Propagation*, vol. 27, no. 5, pp. 640–646, 1979.

[BAL 05] BALANIS C., *Antenna Theory: Analysis and Design*, 3rd ed., Wiley-Interscience, Hoboken, 2005.

[BAU 06] BAUM C.E., "Combining Polarimetry with SEM in Radar Backscattering for Target Identification", *3rd International Conference on Ultrawideband and Ultrashort Impulse Signals*, pp. 11–14, 2006.

[BAU 80] BAUM C.E., SINGARAJU B.K., "The singularity and eigenmode expansion methods with application to equivalent circuits and related topics", *Acoustic, Electromagnetic and Elastic Wave Scattering – Focus on the T-Matrix Approach*, pp. 431–452, 1980.

[BAU 91] BAUM C.E., ROTHWELL E.J., CHEN K.M. *et al.*, "The singularity expansion method and its application to target identification", *Proceedings of the IEEE*, vol. 79, no. 10, pp. 1481–1492, 1991.

[BLI 11] BLISCHAK A.T., MANTEGHI M., "Embedded Singularity Chipless RFID Tags", *IEEE Transactions on Antennas and Propagation*, vol. 59, no. 11, pp. 3961–3968, 2011.

[BOE 81] BOERNER W.M., EL-ARINI M., CHAN C.-Y. *et. al.*, "Polarization dependence in electromagnetic inverse problems", *IEEE Transactions on Antennas and Propagation*, vol. 29, no. 2, pp. 262–271, 1981.

[BRA 30] BRARD E., Process for radiotelegraphic or radiotelephonic communication, Patent US1744036 A, 1930.

[BRO 99] BROWN L., *A Radar History of World War II: Technical and Military Imperatives*, 1st ed., Institute of Physics Publishing, Bristol, 1999.

[BUL 10] BULATSYK O., KATSENELENBAUM B.Z., TOPOLYUK Y.P. *et al.*, *Phase Optimization Problems: Applications in Wave Field Theory*, John Wiley & Sons, New York, 2010.

[CHA 11] CHAABANE H., PERRET E., TEDJINI S., "A Methodology for the Design of Frequency and Environment Robust UHF RFID Tags", *IEEE Transactions on Antennas and Propagation*, vol. 59, no. 9, pp. 3436–3441, 2011.

[CHE 06] CHEN C.-F., HUANG T.-Y., CHOU C.-P. *et al.*, "Microstrip diplexers design with common resonator sections for compact size, but high isolation", *IEEE Transactions on Microwave Theory and Techniques*, vol. 54, no. 5, pp. 1945–1952, 2006.

[CLO 96] CLOUDE S.R., POTTIER E., "A review of target decomposition theorems in radar polarimetry", *IEEE Transactions on Geoscience and Remote Sensing*, vol. 34, no. 2, pp. 498–518, 1996.

[COL 01] COLLIN R.E., *Foundations for Microwave Engineering*, Wiley, New York, 2001.

[COL 12] COLTON D., KRESS R., *Inverse Acoustic and Electromagnetic Scattering Theory*, Springer Science & Business Media, Berlin, 2012.

[COO 14] COOK B.S. *et al.*, "RFID-Based Sensors for Zero-Power Autonomous Wireless Sensor Networks", *IEEE Sensors Journal*, vol. 14, no. 8, pp. 2419–2431, 2014.

[COS 13] COSTA F., GENOVESI S., MONORCHIO A., "A Chipless RFID Based on Multiresonant High-Impedance Surfaces", *IEEE Transactions on Microwave Theory and Techniques*, vol. 61, no. 1, pp. 146–153, 2013.

[COS 15] COSTA F., GENOVESI S., MONORCHIO A. *et al.*, "A Robust Differential-Amplitude Codification for Chipless RFID", *IEEE Microwave and Wireless Components Letters*, vol. 25, no. 12, pp. 832–834, 2015.

[DAS 10] DAS R., HARROP P., RFID Forecasts, Players and Opportunities 2011–2021, IDTechEx, 2010.

[ELA 15] EL-AWAMRY A., KHALIEL M., FAWKY A. *et al.*, "Novel notch modulation algorithm for enhancing the chipless RFID tags coding capacity", *IEEE International Conference on RFID (RFID)*, pp. 25–31, 2015.

[FEN 15a] FENG C., ZHANG W., LI L. *et al.*, "Angle-Based Chipless RFID Tag With High Capacity and Insensitivity to Polarization", *IEEE Transactions on Antennas and Propagation*, vol. 63, no. 4, pp. 1789–1797, 2015.

[FEN 15b] FENG Y., XIE L., CHEN Q. *et al.*, "Low-Cost Printed Chipless RFID Humidity Sensor Tag for Intelligent Packaging", *IEEE Sensors Journal*, vol. 15, no.6, pp. 3201–3208, 2015.

[GAR 15] GARBATI M., SIRAGUSA R., PERRET E. *et al.*, "Low cost low sampling noise UWB Chipless RFID reader", *IEEE MTT-S International Microwave Symposium*, pp. 1–4, 2015.

[GAR 16] GARBATI M., RAMOS A., SIRAGUSA R. *et al.*, "Chipless RFID reading system independent of polarization" *IEEE MTT-S International Microwave Symposium (IMS)*, pp. 1–3, 2016.

[GIR 12a] GIRBAU D., RAMOS A., LAZARO A. *et al.*, "Passive Wireless Temperature Sensor Based on Time-Coded UWB Chipless RFID Tags", *IEEE Transactions on Microwave Theory and Techniques*, vol. 60, no. 11, pp. 3623–3632, 2012.

[GIR 12b] GIRBAU D., LAZARO A., RAMOS A., "Time-coded chipless RFID tags: Design, characterization and application", *IEEE International Conference on RFID-Technologies and Applications (RFID-TA)*, pp. 12–17, 2012.

[GOV 95] GOV S., SHTRIKMAN S., "On isotropic scattering of electromagnetic radiation", *18th Convention of Electrical and Electronics Engineers in Israel*, p. 2.4.4/1-2.4.4/5, 1995.

[GUP 96] GUPTA K.C., GARG R., BAHL I. *et al.*, "Coplanar lines: Coplanar waveguide and coplanar strip", in GARG R., BAHL I., BOZZI M. (eds), *Microstrip Lines and Slotlines*, 2nd ed., Artech House Publishers, London, 1996.

[GUP 11] GUPTA S., NIKFAL B., CALOZ C., "Chipless RFID System Based on Group Delay Engineered Dispersive Delay Structures", *IEEE Antennas and Wireless Propagation Letters*, vol. 10, pp. 1366–1368, 2011.

[HAD 14] HADAMARD J., Lectures on Cauchy's Problem in Linear Partial Differential Equations, Courier Corporation, North Chelmsford, 2014.

[HAN 89] HANSEN R.C., "Relationships between antennas as scatterers and as radiators", *Proceedings of the IEEE*, vol. 77, no. 5, pp. 659–662, 1989.

[HAN 90] HANSEN R.C., "Antenna mode and structural mode RCS: DIPOLE", *Microwave and Optical Technology Letters*, vol. 3, no. 1, pp. 6–10, 1990.

[HAR 63] HARRINGTON R., "Electromagnetic scattering by antennas", *IEEE Transactions on Antennas and Propagation*, vol. 11, no. 5, pp. 595–596, 1963.

[HAR 64] HARRINGTON R.F., "Theory of loaded scatterers", *Proceedings of the Institution of Electrical Engineers*, vol. 111, no. 4, pp. 617–623, 1964.

[HAR 02] HARTMANN C.S., "A global SAW ID tag with large data capacity", *IEEE Ultrasonics Symposium*, 2002.

[HAR 10] HARROP P., DAS R., Printed and Chipless RFID Forecasts, Technologies & Players 2011–2021, IDTechEx, 2010.

[HU 10] HU S., ZHOU Y., LAW C.L. *et al.*, "Study of a Uniplanar Monopole Antenna for Passive Chipless UWB-RFID Localization System", *IEEE Transactions on Antennas and Propagation*, vol. 58, no. 2, pp. 271–278, 2010.

[HUY 70] HUYNEN J.R., Phenomenological theory of radar targets, PhD Thesis, Delft University of Technology, 1970.

[ISL 15] ISLAM M.A., KARMAKAR N.C., "Compact Printable Chipless RFID Systems", *IEEE Transactions on Microwave Theory and Techniques*, vol. 63, no. 11, pp. 3785–3793, 2015.

[JAC 09] JACKSON D.R., "Microstrip Antennas", in VOLAKIS J. (ed.), *Antenna Engineering Handbook*, 4th ed., McGraw-Hill Professional, New York, 2009.

[JAL 05] JALALY I., ROBERTSON I.D., "RF barcodes using multiple frequency bands", *IEEE MTT-S International Microwave Symposium Digest*, pp. 139-142, 2005.

[KAT 13] KATSENELENBAUM B.Z. *et al.*, "Phase optimization problems in antenna synthesis theory", *IX International Conference on Antenna Theory and Techniques (ICATT)*, pp. 22–27, 2013.

[KEN 52] KENNAUGH E.M., Polarization properties of radar reflections, PhD Thesis, Ohio State University, 1952.

[KHA 15] KHALIEL M., EL-AWAMRY A., FAWKY A. *et al.*, "A novel co/cross-polarizing chipless RFID tags for high coding capacity and robust detection", *IEEE International Symposium on Antennas and Propagation USNC/URSI National Radio Science Meeting*, pp. 159–160, 2015.

[KIM 13a] KIM S., TENTZERIS M. M., TRAILLE A. *et al.*, "A dual-band retrodirective reflector array on paper utilizing Substrate Integrated Waveguide (SIW) and inkjet printing Technologies for Chipless RFID Tag and Sensor Applications", *IEEE Antennas and Propagation Society International Symposium (APSURSI)*, pp. 2301–2302, 2013.

[KIM 13b] KIM S., COOPER J., TENTZERIS M.M. *et al.*, "A novel inkjet-printed chipless RFID-based passive fluid sensor platform", *IEEE SENSORS*, pp. 1–4, 2013.

[KIM 13c] KIM S. *et al.*, "A novel dual-band retro-directive reflector array on paper utilizing Substrate Integrated Waveguide (SIW) and inkjet printing technologies for chipless RFID tag and sensor applications", *Microwave Symposium Digest (IMS), IEEE MTT-S International*, pp. 1–4, 2013.

[KNO 04] KNOTT E.F., SHAEFFER J., TULEY M., *Radar Cross Section*, 2nd ed., SciTech Publishing, Raleigh, 2004.

[KOE 75] KOELLE A.R., DEPP S.W., FREYMAN R.W., "Short-range radio-telemetry for electronic identification, using modulated RF backscatter", *Proceedings of the IEEE*, vol. 63, no. 8, pp. 1260–1261, 1975.

[LAZ 16] LAZARO A., RAMOS A., GIRBAU D. *et al.*, "Signal Processing Techniques for Chipless UWB RFID Thermal Threshold Detector Detection", *IEEE Antennas and Wireless Propagation Letters*, vol. 15, pp. 618–621, 2016.

[LEV 85] LE VINE D.M., SCHNEIDER A., LANG R.H. *et al.*, "Scattering from thin dielectric disks", *IEEE Transactions on Antennas and Propagation*, vol. 33, no. 12, pp. 1410–1413, 1985.

[MAC 89] MACAIGNE A., *La Clé du code-barres*, A. Macaigne, Paris, 1989.

[MAC 14] MACHAC J., POLIVKA M., "Influence of mutual coupling on performance of small scatterers for chipless RFID tags", in *Radioelektronika (RADIOELEKTRONIKA), 24th International Conference*, pp. 1–4, 2014.

[NAI 11] NAIR R., PERRET E., TEDJINI S., "Chipless RFID based on group delay encoding", *IEEE International Conference on RFID-Technologies and Applications (RFID-TA)*, pp. 214–218, 2011.

[NIJ 12] NIJAS C.M. *et al.*, "Chipless RFID Tag Using Multiple Microstrip Open Stub Resonators", *IEEE Transactions on Antennas and Propagation*, vol. 60, no. 9, pp. 4429–4432, 2012.

[NIJ 14] NIJAS C.M. *et al.*, "Low-Cost Multiple-Bit Encoded Chipless RFID Tag Using Stepped Impedance Resonator", *IEEE Transactions on Antennas and Propagation*, vol. 62, no. 9, pp. 4762–4770, 2014.

[PAL 91] PALMER R.C., *The Bar Code Book: Reading, Printing, and Specification of Bar Code Symbols*, Helmers Publishing, Dublin, 1991.

[PER 14] PERRET E., *Radio Frequency Identification and Sensors: From RFID to Chipless RFID*, ISTE Ltd, London and John Wiley & Sons, New York, 2014.

[PLE 10] PLESSKY V.P., REINDL L.M., "Review on SAW RFID tags", *IEEE Transactions on Ultrasonics, Ferroelectrics and Frequency Control*, vol. 57, no. 3, pp. 654–668, 2010.

[PÖP 16a] PÖPPERL M., PARR A., MANDEL C. *et al.*, "Potential and Practical Limits of Time-Domain Reflectometry Chipless RFID", *IEEE Transactions on Microwave Theory and Techniques*, vol. 64, no. 9, pp. 2968–2976, 2016.

[PÖP 16b] PÖPPERL M., ADAMETZ J., VOSSIEK M., "Polarimetric Radar Barcode: A Novel Chipless RFID Concept With High Data Capacity and Ultimate Tag Robustness", *IEEE Transactions on Microwave Theory and Techniques*, vol. 64, no. 99, pp. 1–9, 2016.

[PRE 08] PRERADOVIC S., BALBIN I., KARMAKAR N.C. *et al.*, "A Novel Chipless RFID System Based on Planar Multiresonators for Barcode Replacement", *IEEE International Conference on RFID*, pp. 289–296, 2008.

[PRE 09a] PRERADOVIC S., KARMAKAR N.C., "Design of fully printable planar chipless RFID transponder with 35-bit data capacity", *European Microwave Conference*, pp. 013–016, 2009.

[PRE 09b] PRERADOVIC S., ROY S., KARMAKAR N.C., "Fully printable multi-bit chipless RFID transponder on flexible laminate", *Asia Pacific Microwave Conference*, pp. 2371–2374, 2009.

[PRE 09c] PRERADOVIC S., BALBIN I., KARMAKAR N.C. *et al.*, "Multiresonator-Based Chipless RFID System for Low-Cost Item Tracking", *IEEE Transactions on Microwave Theory and Techniques*, vol. 57, no. 5, pp. 1411–1419, 2009.

[PRE 11] PRERADOVIC S., KARMAKAR N.C., AMIN E.M., "Chipless RFID tag with integrated resistive and capacitive sensors", *Asia-Pacific Microwave Conference Proceedings (APMC)*, pp. 1354–1357, 2011.

[PRE 12] PRERADOVIC S., KARMAKAR N.C., *Multiresonator-Based Chipless RFID: Barcode of the Future*, Springer Science & Business Media, Berlin, 2012.

[RAM 11] RAMOS A., LAZARO A., GIRBAU D. *et al.*, "Time-Domain Measurement of Time-Coded UWB Chipless RFID Tags", *Progress in Electromagnetics Research*, vol. 116, pp. 313–331, 2011.

[RAM 12] RAMOS A., GIRBAU D., LAZARO A., "Influence of materials in time-coded chipless RFID tags characterized using a low-cost UWB reader", *Microwave Conference (EuMC), 2012 42nd European*, pp. 526–529, 2012.

[RAM 16a] RAMOS A., LAZARO A., GIRBAU D. *et al.*, *RFID and Wireless Sensors using Ultra-Wideband Technology*, ISTE Press, London and Elsevier, Oxford, 2016.

[RAM 16b] RAMOS A., PERRET E., RANCE O. *et al.*, "Temporal Separation Detection for Chipless Depolarizing Frequency-Coded RFID", *IEEE Transactions on Microwave Theory and Techniques*, vol. 64, no. 7, pp. 2326–2337, 2016.

[RAN 15] RANCE O., SIRAGUSA R., LEMAÎTRE-AUGER P. *et al.*, "RCS magnitude coding for chipless RFID based on depolarizing tag", *Microwave Symposium (IMS), IEEE MTT-S International*, pp. 1–4, 2015.

[RAN 16a] RANCE O., SIRAGUSA R., LEMAÎTRE-AUGER P. *et al.*, "Toward RCS Magnitude Level Coding for Chipless RFID", *IEEE Transactions on Microwave Theory and Techniques*, vol. 64, no. 7, pp. 2315–2325, 2016.

[RAN 16b] RANCE O., SIRAGUSA R., LEMAÎTRE-AUGER P. *et al.*, "Contactless Characterization of Coplanar Stripline Discontinuities by RCS Measurement", *IEEE Transactions on Antennas and Propagation*, vol. PP, no. 99, p. 1, 2016.

[REZ 14a] REZAIESARLAK R., MANTEGHI M., "A Space -Time -Frequency Anticollision Algorithm for Identifying Chipless RFID Tags", *IEEE Transactions on Antennas and Propagation.*, vol. 62, no. 3, pp. 1425–1432, 2014.

[REZ 14b] REZAIESARLAK R., MANTEGHI M., "Complex-Natural-Resonance-Based Design of Chipless RFID Tag for High-Density Data", *IEEE Transactions on Antennas and Propagation*, vol. 62, no. 2, pp. 898–904, 2014.

[REZ 15a] REZAIESARLAK R., MANTEGHI M., *Chipless RFID - Design Procedure and Detection Techniques*, Springer, Berlin, 2015.

[REZ 15b] REZAIESARLAK R., MANTEGHI M., "Design of Chipless RFID Tags Based on Characteristic Mode Theory (CMT)", *IEEE Transactions on Antennas and Propagation*, vol. 63, no. 2, pp. 711–718, 2015.

[RIE 89] RIEGGER S., WIESBECK W., "Wide-band polarimetry and complex radar cross section signatures", *Proceedings of IEEE*, vol. 77, no. 5, pp. 649–658, 1989.

[SHA 13] SHAO B., AMIN Y., CHEN Q. *et al.*, "Directly Printed Packaging-Paper-Based Chipless RFID Tag With Coplanar Resonator", *IEEE Antennas and Wireless Propagation Letters*, vol. 12, pp. 325–328, 2013.

[SHR 09] SHRESTHA S., BALACHANDRAN M., AGARWAL M. *et al.*, "A Chipless RFID Sensor System for Cyber Centric Monitoring Applications", *IEEE Transactions on Microwave Theory and Techniques.*, vol. 57, no. 5, pp. 1303–1309, 2009.

[SIN 50] SINCLAIR G., "The Transmission and Reception of Elliptically Polarized Waves", *Proceedings of the IRE*, vol. 38, no. 2, pp. 148–151, 1950.

[SKO 08] SKOLNIK M.I., *Radar Handbook*, 3rd ed., McGraw-Hill Professional, New York, 2008.

[STO 48] STOCKMAN H., "Communication by Means of Reflected Power", *Proc. IRE*, vol. 36, no. 10, pp. 1196–1204, 1948.

[TAI 05] TAIT P., *Introduction to Radar Target Recognition*, IET, Stevenage, 2005.

[VEN 11] VENA A., PERRET E., TEDJINI S., "Chipless RFID Tag Using Hybrid Coding Technique", *IEEE Transactions on Microwave Theory and Techniques*, vol. 59, no. 12, pp. 3356–3364, 2011.

[VEN 12a] VENA A., PERRET E., TEDJINI S., "A Fully Printable Chipless RFID Tag With Detuning Correction Technique", *IEEE Microwave and Wireless Components Letters*, vol. 22, no. 4, pp. 209–211, 2012.

[VEN 12b] VENA A., PERRET E., TEDJINI S., "Design of Compact and Auto-Compensated Single-Layer Chipless RFID Tag", *IEEE Transactions on Microwave Theory and Techniques*, vol. 60, no. 9, pp. 2913–2924, 2012.

[VEN 12c] VENA A., PERRET E., TEDJINI S., "High-Capacity Chipless RFID Tag Insensitive to the Polarization", *IEEE Transactions on Antennas and Propagation*, vol. 60, no. 10, pp. 4509–4515, 2012.

[VEN 12d] VENA A., Contribution au développement de la technologie RFID sans puce à haute capacité de codage, PhD Thesis, University of Grenoble, 2012.

[VEN 13a] VENA A., BABAR A.A., SYDANHEIMO L. *et al.*, "A Novel Near-Transparent ASK-Reconfigurable Inkjet-Printed Chipless RFID Tag", *IEEE Antennas and Wireless Propagation Letters*, vol. 12, pp. 753–756, 2013.

[VEN 13b] VENA A., PERRET E., TEDJINI S., "Design rules for chipless RFID tags based on multiple scatterers", *Ann. Telecommun.*, vol. 68, no. 7–8, pp. 361–374, 2013.

[VEN 13c] VENA A., PERRET E., TEDJINI S., "A Depolarizing Chipless RFID Tag for Robust Detection and Its FCC Compliant UWB Reading System", *IEEE Transactions on Microwave Theory and Techniques*, vol. 61, no. 8, pp. 2982–2994, 2013.

[VEN 13d] VENA A. *et al.*, "Design of Chipless RFID Tags Printed on Paper by Flexography", *IEEE Transactions on Antennas and Propagation*, vol. 61, no. 12, pp. 5868–5877, 2013.

[VEN 15] VENA A., SYDÄNHEIMO L., TENTZERIS M.M. *et al.*, "A Fully Inkjet-Printed Wireless and Chipless Sensor for CO2 and Temperature Detection", *IEEE Sens. J.*, vol. 15, no. 1, pp. 89–99, 2015.

[VEN 16] VENA A., PERRET E., TEDJINI S. *et al.*, *Chipless RFID based on RF Encoding Particle: Realization, Coding and Reading System*, ISTE Press, London and Elsevier, Oxford, 2016.

[WIE 91a] WIESBECK W., RIEGGER S., "A complete error model for free space polarimetric measurements", *IEEE Transactions on Antennas and Propagation*, vol. 39, no. 8, pp. 1105–1111, 1991.

[WIE 91b] WIESBECK W., KAHNY D., "Single reference, three target calibration and error correction for monostatic, polarimetric free space measurements", *Proceedings of the IEEE*, vol. 79, no. 10, pp. 1551–1558, 1991.

[WIE 98] WIESBECK W., HEIDRICH E., "Wide-band multiport antenna characterization by polarimetric RCS measurements", *IEEE Transactions on Antennas and Propagation*, vol. 46, no. 3, pp. 341–350, 1998.

[WOO 52] WOODLAND N.J, SILVER B., Classifying apparatus and method, Patent US2612994 (A), 1952.

[YAG 86] YAGHJIAN A.D., "An overview of near-field antenna measurements", *IEEE Transactions on Antennas and Propagation.*, vol. 34, no. 1, pp. 30–45, 1986.

[ZHA 06] ZHANG L., RODRIGUEZ S., TENHUNEN H. *et al.*, "An innovative fully printable RFID technology based on high speed time-domain reflections", *Conference on High Density Microsystem Design and Packaging and Component Failure Analysis*, pp. 166–170, 2006.

[ZHE 08] ZHENG L., RODRIGUEZ S., ZHANG L. *et al.*, "Design and implementation of a fully reconfigurable chipless RFID tag using Inkjet printing technology", *2008 IEEE International Symposium on Circuits and Systems*, pp. 1524–1527, 2008.

[ZOM 15] ZOMORRODI M., KARMAKAR N.C., "I Chipless RFID Reader: Low-cost wideband printed dipole array antenna", *IEEE Antennas and Propagation Magazine*, vol. 57, no. 5, pp. 18–29, 2015.

Index